Advance Praise for
The Diabetes LIFEMAP

"*The Diabetes LIFEMAP* captures information about social determinants of disease and home glucose data in a cloud-based diabetes management platform. The LIFEMAP transforms the traditional office-based model of care into a 21st century model that can be used by the entire healthcare team. To enroll in the LIFEMAP all you need is a smartphone and your doctor's participation. Now it is possible to bridge the gap for patient and healthcare provider to co-manage blood sugar in real time between office visits. The LIFEMAP meshes the essential patient care visit with a mechanism for active management of blood sugar using a conventional home glucose meter or more advanced continuous glucose monitor. As data populates in the LIFEMAP, the healthcare team can actively manage your blood sugar and keep you 'on track'".

The Diabetes LIFEMAP can be read from cover to cover or scanned for a relevant case that is similar to one of your patients with diabetes or your own personal journey with the disease. For the healthcare team, Dr. Bleich provides a deep-dive into the mechanisms of diabetes using evidence-based foundation stones that led to the conception of the LIFE-MAP. *The Diabetes LIFEMAP: Data-driven Diabetes Care for the 21st Century* is a winner. For a new generation of medical students and residents, *The Diabetes LIFEMAP* provides a new approach to delivering excellent care with 21st century smartphone technology that is used worldwide. This book provides a road-forward for an epidemic disease that is challenging healthcare systems with excess cost, poor outcomes, and lack of access to care."

Gwendolyn M. Mahon, M.Sc., Ph.D., Dean and Professor,
Rutgers School of Health Professions

"Diabetes is an extraordinarily demanding disease that lasts a lifetime and is impacted by each decision a person living with diabetes makes each day. We have an impressive evidence base demonstrating that effective management of diabetes improves all outcomes including premature mortality, cardiovascular disease, quality of life, and complications such as blindness, amputation and renal failure. A major barrier to enabling people with diabetes to achieve optimal outcomes is in effective day to day support for patients and providers. This book outlines an innovative strategy developed in a resource limited clinical population.

The Diabetes LIFEMAP is a unique strategy to optimize day-to-day decision making for individuals with diabetes. It captures real-time home glucose monitoring and social determinants of disease that can unravel the best care plan. Diabetes is such a challenging problem, particularly for people requiring insulin, that innovation with pragmatic tools accessible to individuals in all health care systems is essential. Individuals living with diabetes need providers to be equipped with many different strategies to help them optimize their care including telemedicine-based approaches. Dr Bleich's patients have experienced great success with the LIFEMAP strategy. Pragmatic implementation studies to define the impact of this and other innovative real life support tools are particularly crucial for the 20% of the 30,000,000 people in the United States living with diabetes and on insulin. I applaud the creativity and practicality of the LIFEMAP strategy and look forward to data in its effectiveness beyond Dr Bleich's clinical setting."

Jane EB Reusch, MD ASCI AAP, Professor of Medicine, Bioengineering and Biochemistry Endocrinology, Metabolism and Diabetes. Associate Director, Center for Women's Health Research University of Colorado Anschutz Medical Campus SOM. VA Staff Physician and Merit Investigator, CCTSI-VA Liaison and Research Studio Program Director

The Diabetes LIFEMAP

The
Diabetes
lifemap

*Data-Driven Diabetes Care
for the 21st Century*

DAVID BLEICH, MD

NEW YORK

LONDON • NASHVILLE • MELBOURNE • VANCOUVER

The Diabetes LIFEMAP

Data-Driven Diabetes Care for the 21st Century

Published in New York, New York, by Morgan James Publishing. Morgan James is a trademark of Morgan James, LLC. www.MorganJamesPublishing.com

ISBN 9781642799064 paperback
ISBN 9781642799071 eBook
Library of Congress Control Number: 2019918125

Cover & Interior Design by:
Christopher Kirk
www.GFSstudio.com

Cover Illustration by:
Alexandra Palatucci

Morgan James is a proud partner of Habitat for Humanity Peninsula and Greater Williamsburg. Partners in building since 2006.

Get involved today! Visit
MorganJamesPublishing.com/giving-back

Dedication

This book is dedicated to many people in my life. My wife Lisa comes first because she is the driving force in my life who always guides me down the right path. My father Abe would have been proud but would not have said too much. My mother-in-law Tanis would have bragged to anyone willing to listen. My mother Rhoda always believes in me, and my father-in-law Dick would share his excitement openly.

I am indebted to my medical colleagues, my chairman Dr. Marc Klapholz, and Dean Robert Johnson who have entrusted me to oversee the Division of Endocrinology, Diabetes, & Metabolism at Rutgers New Jersey Medical School for the past 15 years.

Acknowledgments

I didn't intend to write this book. Perhaps that is a good thing. Over the last decade I had struggled with treating my patients with diabetes. During that time, I didn't know what to do or how to approach their myriad of complex medical issues. This is not to say I practiced medicine differently from my colleagues and peers. In fact, my practice followed all the standards of care and treatment options for diabetes, but somehow things didn't connect for many of my patients. There were some instances when a patient was easy to manage and there were no unresolved problems. At other times I felt like I was under siege: a patient was too complex, there wasn't enough time, or there were too many roadblocks to care. I realized I needed a solid, structured approach to deliver better care for complex patients with diabetes.

The LIFEMAP evolved as a logical extension of my need to organize and prioritize patient care. I came to realize the LIFEMAP wasn't just for complex patients. Why? Well, because all patients with diabetes have complexity in their own personal way. So, thank you to all my patients in Newark and those who sought me out in Newark. You have guided me, supported me, and didn't give up on me as your doctor. You brought me to the LIFEMAP.

One reason for this book is my wife, Lisa. After listening to my endless yammering about this LIFEMAP I was using for my diabetes care, she said, "Why don't you just write a book?"

Okay, so I did. Thank you, sweetie. As usual, you put me on the right path.

To my children: Rebecca, Gabby, and Kayla. Your annoying father has done something good. Thank you for your support along the way.

My mother, Rhoda, is my own rooting gallery. She is always there for me and I know she is proud of this effort. Thanks, Mom.

My colleagues at Rutgers New Jersey Medical School and Rutgers University do amazing things on a daily basis. I am fortunate to work in an environment of scholarship, excellence, and decency. It is hard to live up to the high standards and achievements of such individuals, but they make me try harder to achieve something great. Specifically, I would like to thank my boss, Dr. Marc Klapholz, Chairman of Medicine and tenured professor who has supported my efforts and let me follow my passion.

I have worked closely with Bob Gold, Jane Hindes-Miller, and Gary Pollack to build and implement the Diabetes LIFEMAP as a cloud-based, personalized diabetes care platform. They are outstanding collaborators and have taken the LIFEMAP from concept to reality. We look forward to spreading the LIFEMAP worldwide throughout healthcare systems to improve the lives of individuals with diabetes and to overcome roadblocks in care faced by all healthcare providers today.

Lastly, I thank my editor Justin Spizman, best-selling author and Georgia author of the year, for his guidance and direction in elevating this book to his high standards.

Table of Contents

Foreword

The use of new technology to better control diabetes mellitus is increasing exponentially in all age groups. Although there are several treatment options available, many people find managing the condition challenging. New technology could help reduce this burden.

The Diabetes LIFEMAP takes a real world view of this complex disease and boils it down to essential elements for successful blood sugar control. By capturing both social determinants of disease and home glucose data in a cloud-based diabetes management platform, the LIFEMAP transforms the office-based model of care we have used for the past 50 years into a 21st century model that is personalized, convenient, and efficient. Now it is possible for patient and healthcare provider to co-manage blood sugar in real time without face-to-face interactions that limits the care process. The LIFEMAP meshes the essential patient care visit with a mechanism for active management of blood sugar using a conventional home glucose meter or more advanced continuous glucose monitor. As data accrues in the LIFEMAP, the healthcare team can actively participate in improving blood sugar and integrate nutritional advice and behavioral prompts to keep the patient "on track". Over time, the LIFEMAP becomes a habit through gentle text prompts that remind patients of their times for blood sugar checks and allows the patient to see their progress whenever they want through a cloud-based patient portal.

The Diabetes LIFEMAP is an easy read with information for both diabetes patient and healthcare provider. Real world case presentations give the patient an easy way to see the LIFEMAP in motion and how a simple set of rules for home glucose monitoring can lead to outstanding outcomes. For healthcare providers, Dr. Bleich provides a deep-dive into the dynamics of insulin secretion and glucose homeostasis using evidence-based foundation stones that led to the conception of the LIFEMAP. For both diabetes patient and healthcare provide *The Diabetes LIFEMAP: Data-driven Diabetes Care for the 21st Century* is a must read. This book *is innovative, clear, and able to open pathways to new ideas;* a real game-changer for a world that is challenged with an epidemic of diabetes that continues to increase.

Massimo Pietropaolo, MD, McNair Scholar. Professor of Medicine & Immunology, Diabetes Research Center, Baylor College of Medicine

Chapter 1

Changing the Game

A basic tenant of scientific research is to create an experiment with tightly controlled variables, and then stress one variable in the system to identify potential defects or anomalies. We see this practice create remarkable discoveries that can assist us in making tremendous strides in modern medicine. Sometimes you don't exactly plan for the experiment, or even determine the controlled variables at first. They come to you as a function of what might be occurring in the world around you. That is what happened to me and proved to be the greatest experiment I have ever (and unintentionally) conducted.

Moving my practice to the inner city changed me. As a physician with knowledge about diabetes on a molecular and clinical level, I realized that my practice of diabetes was misaligned with the needs of my patients. But my work in a resource-constrained environment allowed me to prioritize certain critical elements of diabetes care from those that aren't all that important, to the ones that could be deemed as lifesaving. Of course, I had to pivot along the way.

The transition in my practice of diabetes did not occur through an epiphany, but rather evolved in the day-to-day necessity of trying to figure out an optimal approach to deliver high-quality care to my patients. I was dealing with extremely sick patients, suffering from a lack of nutrients, healthy practices, and at times lack of access to quality medical care. It was sad to see this decline from a healthy lifestyle. But as a doctor, you don't get to pick and choose ideal circumstances or environments. My responsibility is to help anyone and everyone that needs it, no matter how challenging the diagnosis or treatment might seem.

It was quite apparent that I would not have all the resources I needed when I first moved from a high-profile healthcare center at City of Hope National Medical Center in Duarte, California to Rutgers New Jersey Medical School and University Hospital in Newark, New Jersey. My diabetes patients in Newark were infinitely more complex than those on the West Coast, with long lists of medications and complications that felt mind numbing at times. Many of my patients did not have resources to buy medicine or pay for healthcare insurance, let alone eat on a regular basis. They lacked access to quality food, and there were no parks or other areas where they could exercise. The playgrounds were often overrun with gang activity, and it simply wasn't safe to be there at any point in the day. It was a sobering equation, and many of my patients reported they'd stay in their homes just to be safe from the violent lifestyle around them. Some had literacy problems and others had language barriers that prevented them from social advantage. As a physician, it felt almost impossible to determine a starting point of attack to help my patients. I was overwhelmed and lost in the myriad medical and social problems that are part of inner-city life.

At first, I fell back on old habits of my medical training that emphasized meeting with patients to learn more about their history, offer them a physical exam, a medical assessment, and then finalize our time together with a thorough treatment plan. It had worked in California, so why wouldn't it work here? After a while, it was clear I was getting lost in the complexities of my patients, and at times could not accomplish even a single goal of care for any one of them. I knew I had to make a change, not for me but for their health. This proved to be the easy part. The challenging piece was figuring out what this change might actually look like. Clearly, I could not simultaneously manage all the problems of each patient. However, most of my patients were coming to see me for a common reason: to get control of their blood sugar. Knowing that, I at least felt like there was an obvious starting point.

I recognized many of my patients had multiple layers of complexity to their medical care and their lives, and I felt I had to develop a methodology for engaging my patients in their diabetes care. This was a small yet powerful revelation, which led me to break down the components of chronic diabetes care. In doing so, I generated and built a priority list of high-, moderate-, and low-impact interventions that I might offer patients in need.

As a physician, it is difficult to admit error. I like to think of myself as an expert and know-it-all. However, it was sobering and informative to recognize I was part of the environment offering my patients less-than-optimal diabetes care. When I looked back at the vast number of patients with diabetes I believed I had treated well, I realized my approach to diabetes care lacked thought and rigor, even in California.

I was participating in a crash-course curriculum, and it is now apparent to me that the standard rubric for patient care taught in all medical schools throughout the country (e.g., History, Physical Exam, Assessment, and Plan) is flawed because the care model is derived from the acute hospital setting where it thrives. However, chronic care in the outpatient setting is entirely different. To start, you can divide it into high-impact, moderate-impact, and low-impact interventions. Not every patient needs the same level of attention, care, or treatment.

In response to this trial by fire, I reconfigured my diabetes care model to place the maximum time element on high-impact interventions, and the minimum time element on low-impact interventions. Since time constraints and limitations are an ongoing challenge for all physicians, I figured this rational approach would help to disrupt the traditional model and deliver quality chronic diabetes care to those with the most need for it. Treat the life-threatening problems first, and then work your way up to those that aren't so serious.

Years later, now my diabetes care is intimate insofar as it builds a treatment plan around the lifestyle and individualized needs of each patient. As we will see later in a chapter entitled "LIFEMAP and Social Determinants of Diabetes: Ground-Level View," I have tried to consolidate the critical elements of the diabetes visit into a single computer screen shot that encapsulates the essential individualized variables that go into making an informed treatment decision for a patient with diabetes and thus optimize care. Funny enough, as doctors we tend to complicate things, while patients respond best to a simplified and highly structured treatment program. The easier I could make it on my patient, the greater the chance he or she would start to adjust his or her behavior and overall health practices, especially where there is limited access to information and resources.

While it would be simplistic to surmise that this approach to diabetes care is only relevant for an underclass, inner-city patient population, I believe the principles outlined in this monograph can and should be applied to all individuals with diabetes (or prediabetes) regardless of socioeconomic status. It matters more when the patient is not set up for success per se, but this notion of overcomplicating treatment protocol is an ailment from which we all suffer.

It is also important to recognize that many institutions have a full range of seamless diabetes services that include physician care, on-site certified diabetes education, nutrition, and social workers. It is my intent to help provide these robust centers with an integrated platform for all healthcare workers that come into contact with diabetic patients. I believe this novel approach to managing diabetes will be as efficient and effective for a center with multiple healthcare providers as it is for a place with limited resources or even a single doctor.

Limited Resources, Big Problems

We had few resources to care for our patients with diabetes when I first came to University Hospital in Newark, New Jersey, in 2004. This included two physicians, one endocrine fellow, and two medical health technicians (one who registered the patients and one who obtained vital signs and capillary glucose). My clinical experience lined up consistently with research studies like the UKPDS[1] randomized clinical trial for type 2 diabetes and the DCCT/EDIC study[2,3] for type 1 diabetes, which showed that hemoglobin A1C improves

1   UK Prospective Diabetes Study (UKPDS) Group, "Intensive blood-glucose control with sulfonylureas or insulin compared with conventional treatment and risk of complications in patients with type 2 diabetes (UKPDS 33)." *The Lancet*, 1998, 352:837-853.

2   Diabetes Control and Complications Trial Research Group, D. M. Nathan, S. Genuth, J. Lachin, P. Cleary, O. Crofford, M. Davis, L. Rand, and C. Siebert, "The Effect of Intensive Treatment of Diabetes on the Development and Progression of Long-Term Complications in Insulin-Dependent Diabetes Mellitus," *New England Journal of Medicine*, 1993, 329:977-986.

3   The Diabetes Control and Complications Trial (DCCT)/Epidemiology of Diabetes Interventions and Complications (EDIC) Research Group, "Effect of intensive diabetes therapy on the progression of diabetic

over time with intensive treatment, but then regresses back to base-lines later on.

However, I learned there is a significant benefit even in "short-term" diabetes control. What is important now is that my care is consistent and my rubric is the same for each patient because I focus my treatment approach on the LIFEMAP, which we will outline in the forthcoming pages. Institutions with a full range of services for individuals with diabetes deliver different perspectives and at times conflicting talking points. The reason this occurs is because each of us has our own particular expertise we impart to our patients and believe that these "words of wisdom" are critical components of the overall diabetes care, thereby leading to excellent glucose control.

However, evidence from the National Health and Nutrition Examination Surveys (NHANES) and other studies clearly demonstrate that significant improvement in glucose control over the decades is still necessary despite advances in diabetes pharmacotherapy.[4] As shown by data from the CDC (2000: 4.4% of population versus 2015 7.4% of population), our collective efforts to slow the rate of diabetes in the US have failed over the past 15 years. Worldwide, our efforts have no greater impact on forestalling or diminishing the rates of diabetes. Therefore, recognizing the increased number of individuals with diabetes, a consistent, structured, personalized approach to diabetes care seems reasonable and appropriate for 21st century medicine. That is the purpose of this book.

The first six chapters of my monograph define the underlying principles that led me to develop the LIFEMAP approach to diabetes care. These

retinopathy in patients with type 1 diabetes: 18 years of follow-up in the DCCT/EDIC," *Diabetes*, 2015, 64:631-642.

4   D. Giugliano, M. I. Maiorino, G. Bellastella, P. Chiodini, A. Ceriello, and K. Esposito, "Efficacy of Insulin Analogs in Achieving the Hemoglobin A1c Target of <7% in Type 2 Diabetes," *Diabetes Care*, 2011, 34:510-517.

chapters describe the integration of basic physiology with clinical care and build foundation stones for the following chapters. In chapter 6, I also discuss the factors in life that make diabetes a challenging disease and can derail even the most carefully thought-out plan. These social determinants of disease are often as much the cause of diabetes as they are the result.

In chapter 7, I give a brief and simplified review of nutrition as it pertains to diabetes management. Chapters 8 through 10 set up the principles for a new chronic care diabetes model and what this might look like for the first three healthcare visits. I describe the "how to" for each of the critical first three visits that establish the LIFEMAP approach for the patient. In chapter 11, I address the "gaps" that arise from using the LIFE-MAP approach and show how the healthcare provider can navigate these gaps. Gaps are defined as situations where there are a limited number of home glucose values that can be used to interrogate and reconfigure the treatment strategy. Navigating the gaps is an important component of using the LIFEMAP because if glucose test strips are the rate-limiting factor for robust data collection, then we need "work-arounds" to optimize the use of data.

Finally, chapters 12 and 13 describe a series of hypothetical cases that commonly occur during chronic diabetes management and how the LIFEMAP is being configured as a cloud-based telemedicine diabetes management system. The intent of these cases is to give the healthcare provider and patient an in-depth view of how to use the LIFEMAP. The LIFEMAP approach to diabetes changes the care delivery from the present physician/healthcare provider "centric" model to a patient-centered, data-driven model. Chapter 14 gives concluding remarks about the future of diabetes care in our mobile, technology-driven society that needs access to care outside the four walls of a medical office and expects high-quality outcomes.

It was a somewhat shocking experience to see sickly patient after patient. These were some of the most complicated cases I've ever worked on, and my patients often suffered from a deadly lack of care, knowledge, and access to a healthy lifestyle. But maybe most important to them is that no one ever took the time to break things down and actually explain to them why they are suffering from diabetes. As I developed the LIFE-MAP approach, I recognized that patient after patient responded to this method, willing to take even a few steps in the right direction. In some ways it wasn't about treating the disease; it was about educating the patient why the disease existed in the first place and recognizing that the social determinants of disease can overshadow the problem with blood sugar.

Doctors and healthcare providers across the world face a similar challenge. But the LIFEMAP approach offers a great deal of hope and benefit for them (who can offer higher quality care to more patients) and the patient (who can work to better control or even forestall their diabetes). Combine those two tremendous advantages, and there is no question LIFEMAP is a true difference maker. But before we dive headfirst into the details, let's spend some time on the actual physiology of insulin and pharmacologic replication, which is the topic of the next chapter.

Chapter 2

Physiology of Insulin and Pharmacologic Replication

I n writing a book on diabetes, a logical and important place to begin our journey is with a simple discussion about insulin. Insulin is a hormone produced in the pancreas by the islets of Langerhans that regulates the amount of glucose in the blood. The islets of Langerhans contain a very specialized cell type called the beta cell, and these little guys are insulin production factories. The lack of insulin (either relative or absolute) causes diabetes, which will be our focus moving forward.

Diabetes is a not a single disease entity (or two disease entities for that matter). Older terms like type 1 diabetes and type 2 diabetes add to confusion for both healthcare providers and patients. Why?

Because in the 21st century we understand that there are many different forms of diabetes that are "dumped" into these two bins.

But the truth is that diabetes is a complex group of diseases that have common elements among all patients, and also specific elements that generate a great diversity of the disease. One common element we see is decreased insulin production in the pancreatic beta cells that leads to elevated blood sugar. Defects in the liver, skeletal muscle, gut, endocrine hormone systems, and nervous system are distinct elements that affect certain individuals, but not others.

Genetic factors combine with environmental inputs (diet, exercise, weight, sleep, and stress) to create an imbalance between the level of blood sugar and the amount of insulin necessary to control the blood sugar. In general, normal blood sugar is tightly regulated between 80 mg/dl and 120 mg/dl. As blood sugar starts to rise mildly, a condition called prediabetes occurs, whereby fasting blood sugar is greater than 100 mg/dl, but less than 126 mg/dl. A large population of individuals was used to define these cut points (>100 and <126), some with "normal" blood sugar and others with elevated blood sugar, thereby reflecting statistical norms of large "healthy" populations rather than absolute benchmarks.

When fasting blood sugar is between 100 mg/dl and 125 mg/dl, we call it impaired fasting glucose, a pre-diabetic condition. Diabetes occurs when fasting blood sugar exceeds 126 mg/dl on two or more occasions. Blood sugar rises under normal conditions after meals, but should not be greater than 140 mg/dl. Impaired glucose tolerance is defined as blood sugar after a meal greater than140 mg/dl and less than 200 mg/dl. This too is a pre-diabetic condition.

Diabetes can be defined as the condition of blood sugar being greater than 200 mg/dl on two or more occasions after a meal. Over time, as the blood sugar gradually rises, it puts stress on pancreatic beta cells to produce more insulin. Failing beta cells set a vicious cycle in motion that leads to high blood sugar and beta cell injury. In turn, this leads to higher blood sugar and more beta cell injury. As the cycle continues, the blood sugar rises and insulin production falls. This directly leads to diabetes. We call the phenomenon "glucose toxicity."

I am often asked if diabetes is a permanent condition. Years ago, my standard answer was once you have diabetes you always have diabetes. But I have been proven incorrect. Long-term studies on obese individuals with type 2 diabetes undergoing bariatric (stomach) surgery demonstrate diabetes reversal in ~30.4% of patients at 15 years of follow-up.[5] The implication of such studies is that when significant weight loss is achieved, then there is reestablishment of an equilibrium between insulin and glucose whereby normal glucose regulation is restored. So, it is possible for certain individuals to reverse or find remission of type 2 diabetes. However, this is not the case for most people. Moreover, the necessary ingredients for reestablishing this critical balance is not well understood at present, but certainly requires significant weight loss that is sustained for many years.

Decreased insulin resistance is one explanation for the normalization of blood sugar after weight loss surgery. Insulin resistance sits on the other side of the insulin and glucose equation. Certain tissues like skeletal muscle, liver, and fat have the ability to respond to insulin. In skeletal muscle this response is increased glucose uptake. In the liver the response is decreased glucose production (the liver is a sugar storage factory). In the fat cells, insulin inhibits the release of fat stores (called lipolysis). Insulin resistance means these critical tissues do not respond to insulin normally; they resist the action of insulin and thereby help to create an environment in the blood of elevated glucose

5   L, Sjostrom, M. Peltonene, P. Jacobson, et al, "Association of bariatric surgery with long-term remission of Type 2 diabetes and with microvascular and macrovascular complications," *JAMA*, 2014, 311(22):2297-2304.

and free fatty acids. This in turn leads to more insulin resistance and its own vicious cycle.

Basal and Bolus Insulin

The concept of basal (low levels of insulin secretion throughout the day in type 1 & 2 diabetes) and bolus insulin (rapid high-level insulin release during meals) replacement therapy is a good starting point in our effort to recapitulate normal pancreatic beta cell physiology and function in an individual with diabetes.

Under conditions of health, pancreatic beta cells have an ample supply of insulin and glucagon to control glucose production from the liver (basal insulin release) and glucose obtained from food sources (bolus insulin release). It is worthwhile mentioning that hepatic (liver) glucose production also contributes significantly to post meal glucose excursions as demonstrated by Rizza and others[6] through a hormone called glucagon (stimulates liver glucose production) as well as insulin (inhibits liver glucose production). So basal insulin is not purely for the sake of the liver glucose production overnight.

Insulin Replacement in Diabetes

It is important to mention that data obtained from the UKPDS study teaches us that about 50% of patients with type 2 diabetes will require insulin in 6 years.[7] Well-controlled blood sugar might be one of the factors that prevents or forestalls this conversion from oral

---

6   R. A. Rizza, "Pathogenesis of fasting and postprandial hyperglycemia in Type 2 diabetes: Implications for Therapy," *Diabetes,* 2010, 59(11):2697-2707.

7   UK Prospective Diabetes Study (UKPDS) Group, "Intensive blood-glucose control with sulfonylureas or insulin compared with conventional treatment and risk of complications in patients with type 2 diabetes (UKPDS 33)." *The Lancet,* 1998, 352:837-853.

agents to insulin. Therefore, controlling blood sugar might not only be important for preventing eye, kidney, and nerve disease associated with diabetes. The hierarchy of insulin replacement for diabetes based on recapitulation of healthy beta cell physiology can be summarized below:

1. Real-time Continuous Glucose Monitoring (CGM) + insulin pump
2. Basal-Bolus-Correction dosing on insulin pump minus CGM
3. Basal long-acting insulin + bolus short-acting insulin (4-shot insulin regimen)
4. Premixed insulin before breakfast + short-acting insulin at dinner + bedtime basal insulin (3-shot regimen)
5. Bolus insulin dosing before meals (3-shot insulin regimen)
6. Premixed insulin before breakfast and dinner (2-shot insulin regimen)
7. Bedtime basal long-acting insulin + oral agents (1-shot insulin regimen)

It is also exciting to know that in the near future we should see a bionic pancreas that pumps both insulin and glucagon (a hormone that counteracts insulin and raises the blood sugar), automatically regulating blood sugar.[8] At each level of therapeutic intensity, it might be expected that for diabetic patients with uncontrolled blood sugar a step "up the ladder" will lead to better blood sugar control because you come closer to approximating normal beta cell physiology. It is understood that a patient with hemoglobin A1C of 9.5% on a 2-shot insulin regimen typically can be expected to attain an A1C in the 7.5%-8.5% range by moving to a 3-shot insulin regimen depending on lifestyle compliance.

---

8   S. J. Russell, F. H. El-Khatib, M. Sinha, K. L. Magyar, K. McKeon, L. G. Goergen, C. Balliro, M. A. Hillard, D. M. Nathan, and E. R. Damiano, "Outpatient Glycemic Control with a Bionic Pancreas in Type 1 Diabetes," *New England Journal of Medicine*, 2014, 371:313-325.

Therefore, standard guidelines fall short in describing accurate A1C targets for individual patients.[9]

Incretin Hormones

Another class of critical agents called incretin hormones potentiates glucose-stimulated insulin secretion during meals and incretin deficiency is present early in the development of type 2 diabetes.[10] Incretins are gut hormones that interact with pancreatic beta cells to improve their function. These hormones also have indirect effects that lead to decreased sugar production from the liver and slowed digestion in the stomach. Incretin hormones have been commercialized for years, and are now available to treat individuals with type 2 diabetes. Decreased incretin hormone production is part of a mechanism that leads to post-prandial (meal) hyperinsulinemia (excess insulin secretion overall) and loss of first-phase insulin secretion in some individuals (insulin release is generally divided into a rapid first phase occurring between 1-5 minutes and a prolonged second phase of insulin secretion that can typically last up to 30-60 minutes).[11] As a consequence of insufficient incretin-potentiated glucose-stimulated insulin secretion early on, the beta cell is forced to play catch-up to compensate for transient hyperglycemia during the early stage of feeding. This leads to excess insulin secretion in the second phase of insulin release. Excess glucagon production or lack of insulin suppression of glycogenolysis (glycogen breakdown into glucose) from

9   American Diabetes Association, "Standards of Medical Care in Diabetes," *Diabetes Care,* 2018, 41(Suppl 1), s1-172.

10 M. Nauck, "Reduced incretin effect in type 2 (non-insulin-dependent) diabetes," *Diabetologia,* 1986, 29:46-54.

11 L. A. Scrocchi, T. J. Brown N. Maclusky, et al., "Glucose intolerance but normal satiety in mice with a null mutation in the glucagon-like peptide 1 receptor gene," *Nature Medicine,* 1996, 2:1254-1258.

the liver or both is also a significant component that causes excessive hepatic glucose production and contributes significantly to hyperglycemia both fasting and after meals.[12, 13]

Fortunately, incretin hormone therapy can decrease the hyperglucagonemia that occurs early on with both type 1 diabetes and type 2 diabetes.[14] Some researchers propose that post meal hyperinsulinemia[15] creates a vicious cycle of hyperinsulinemia leading to increased fuel storage and increased appetite, leading to further feeding and hyperinsulinemia until the beta cell can no longer compensate. At this point, overt hyperglycemia ensues and at this stage we call it type 2 diabetes, although the metabolic decompensation was happening over months to years in most cases.[16]

As our understanding of beta cell physiology increases, so does the complexity and intermeshing of these various metabolic cycles because we discover new mechanisms involved in the diabetic process. Conse-

12   P. Shah, A. Vella, A. Basu, et al., "Lack of suppression of glucagon contributes to postprandial hyperglycemia in subjects with type 2 diabetes," *The Journal of Clinical Endocrinology & Metabolism*, 2000, 85(11):4053-4059.

13   G. Bock, C. Dalla Man, M. Campioni, et al., "Pathogenesis of pre-diabetes: mechansims of fasting and postprandial hyperglycemia in people with impaired fasting glucose and/or impaired glucose tolerance," *Diabetes*, 2006, 55(12):3536-3549.

14   E. Naslund, J. Bogefors, S. Skogar, et al., "GLP-1 slows solid gastric emptying and inhibits insulin, glucagon, and PYY release in humans," *American Journal of Physiology*, 1999, 277(3):R910.

15   B. E. Corkey, "Hyperinsulinemia: Cause or Consequence?" *Diabetes*, 2011, 61(1):4-13.

16   S. Lillioja, D. M. Mott, M. Spraul, et al., "Insulin Resistance and Insulin Secretory Dysfunction as Precursors of Non-Insulin Dependent Diabetes Mellitus: Prospective Studies of Pima Indians," *New England Journal of Medicine*, 1993, 329:1988-1992.

quently, picking the correct medication for a specific patient without knowledge of the underlying molecular disease causing the diabetic state becomes more complex.

The Goal of Normal Physiology

It is important to realize when treating a patient with diabetes that all efforts should be focused on restoring the normal physiology as much as possible. This is important because the goal of all diabetes treatment is to restore normal physiologic glucose levels in the blood. In some cases, intensive lifestyle modification is enough to control hemoglobin A1C and glucose without resorting to pharmacological means.[17] Unfortunately, it appears in most cases that intensive lifestyle modification wanes after two years.[18]

Our behaviorists teach us that human beings tend to be creatures of habit and our food preferences are typically stable over time.[19, 20] Therefore, efforts to enforce demanding changes in lifestyle unravel because our natural tendencies, genetic factors, or environmental cues lead us back to our preferred eating habits. When this occurs, patients surrender all the hard work they did to lose weight in the first place. As their behavioral patterns change, they often gain weight back. Since weight control is a key factor in the balance between glucose and insulin, improvements

17 The Diabetes Prevention Program (DPP) Research Group, "The Diabetes Prevention Program (DPP)," *Diabetes Care,* 2002, 25(12):2165-2171.

18 The Diabetes Prevention Program Research Group, "10-year follow-up of diabetes incidence and weight loss in the Diabetes Prevention Program Outcomes Study," *The Lancet,* 2009, 374(9702):1677-1686.

19 N. Lien, L. A. Lytle, and K. I. Klepp, "Stability in consumption of fruit, vetetables, and sugary foods in a cohort from age 14 to age 21," *Preventive Medicine,* 2001, 33(3):217-226.

20 F. C. Cruz, E. Ramos, C. Lopes, and J. Araújo, "Tracking of food and nutrient intake from adolescence into early adulthood," *Nutrition,* 2018, 55-56, 84-90.

in glucose control resulting from weight loss can be easily upended when patients gain back weight. As a general rule, weight loss that occurs in year one of a diet plan is surrendered in year two with regain back to the starting weight over time.

Efforts to normalize the balance between insulin production from the islets of Langerhans in the pancreas and insulin resistance (or sensitivity) in skeletal muscle or other organs have had varying outcomes with both bariatric (stomach) surgery and medications. In the Swedish Obesity Study, subjects were randomized into groups of medical weight loss, gastric banding, gastric sleeve surgery, or full gastric bypass surgery. Those individuals treated with full gastric bypass procedures (but not the other groups) had diabetes reversal rate of ~70% at 2 years and ~30% at 10 years post procedure.[5] Medical therapies without sustained weight loss has some success in lowering hemoglobin A1C, but not for reversing diabetes.

In the ADOPT trial, diabetic patients were randomized to one of three oral medications: glyburide, metformin, or rosiglitazone.[21] Although A1C lowering was greatest at 1 year with glyburide (sulfonylurea), at 5 years rosiglitazone (PPAR-γ drug) provided the most protection (as defined by maintaining hemoglobin A1C less than 8.0%), while metformin was intermediate between the two agents. Unfortunately, controversy surrounding the entire class of PPAR-γ agonists (rosiglitazone and pioglitazone being two such drugs) has led physicians away from prescribing these agents in part because of weight gain associated with treatment. As for PPAR- γ agonists, they appear to have safe cardiovascular outcome profiles and do not show a signal for increased risk of bladder cancer (a rare occurrence nevertheless). We have therefore come to use metformin as a foundation stone drug in the treatment of type

---

21 S. E. Kahn, J. M. Lachin, B. Zinman, S. M. Haffner, R. P. Aftring, G. Paul, B. G. Kravitz, W. H. Herman, G. Viberti, R. R. Holman; ADOPT Study Group, "Effects of rosiglitazone, glyburide, and metformin on β-cell function and insulin sensitivity in ADOPT," *Diabetes*, 2011, 60(5):1552-1560.

2 diabetes, prediabetes, and PCOS (polycystic ovarian syndrome) and appropriately so even though the precise mechanism of action has not yet been fully worked out.[22] Metformin turns out to be a weight-neutral to slight weigh-negative drug, meaning that many individuals will lose three to five pounds by taking metformin.

It is important to remember the composite daily insulin production as shown in figure 1 when attempting to recapitulate normal physiological insulin secretion.

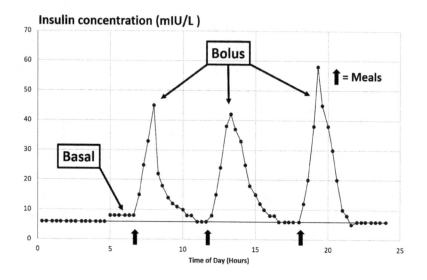

The physiology of the beta cell is often overlooked in the treatment of diabetes because "top-down" approaches to therapy are often effective and require little effort.

For example, a given diabetes drug will lower glucose and hemoglobin A1C by a certain percentage as indicated by randomized clinical studies. The healthcare provider can treat a patient with diabetes drug

---

22   A. K. Madiraju, Y. Qiu, R. J. Perry, Y. Rahimi, and X. M. Zhang, et al.,
      "Metformin inhibits gluconeogenesis via a redox-dependent mechanism in
      vivo," 2018, *Nature Medicine*, 24:1384-1394.

"A" and have some confidence that she will respond according to the package insert. Therefore, we can prescribe a particular drug for a given patient and "see what happens" with knowledge that on average a certain group of patients will respond according to the study results.

This is easy, quick, and does not take much thinking from the healthcare provider. A top-down approach is generally useful in recent onset, easy to manage type 2 diabetes. Unfortunately, a top-down approach to diabetes care runs amuck when the blood sugar control is lost over time. Then, a "bottom-up" approach is necessary and one might argue that it should have been set in place right from the start of care. We will talk more about top-down and bottom-up approaches to diabetes care. A patient with newly diagnosed diabetes might benefit from metformin or a combination of metformin plus DPP-4 inhibitor (a drug class that helps the pancreatic beta cell make more insulin), but the clinician does not typically pay attention to the underlying pathophysiology of the disease. Moreover, we lack adequate biomarkers (blood tests) at present to ascertain with precision many forms of both type 1 and type 2 diabetes.

Recent advances in pharmacotherapy have introduced a new class of glucose-lowering agents called SGLT-2 inhibitors that inhibit resorption of glucose in the proximal kidney tubule and promote glucose lowering with modest weight loss. This class of agents has been used both as monotherapy and in combination with other glucose-lowering drugs. High-fasting plasma glucose as a primary defect is indicative of ineffective suppression of hepatic glucose production overnight. This should lead the physician to consider bedtime insulin early in the treatment cycle to restore "normal" physiology; this would be an example of a bottom-up approach to treatment. The simplest explanation for top-down treatment is the physician has significant time constraint and contemplating disease pathophysiology does not treat the patient, is not compensated, and does not get her out of the office in a timely fashion.

In summary, new clinical and scientific information about diabetes has revealed an amalgamation of complex diseases with many variants that all share a common element of elevated blood sugar. We continue to lump patients generally in the two bins of type 1 diabetes and type 2 diabetes. This is not a productive approach for future knowledge and better care. Unfortunately, we have not reached a point in medicine where we can inform a patient with precision as to what form of diabetes they have.

Obesity and lifestyle are major contributors to the epidemic nature of diabetes throughout the world and newer treatments have improved our ability to strategically target different organ systems (e.g., pancreas, liver, skeletal muscle, gastrointestinal tract) to improve glucose control and overall outcomes. Nevertheless, our lifestyles and environmental inputs seem to be winning out.

As a starting point for 21st century diabetes care, I take the perspective that all efforts should be focused on replicating normal physiology of beta cell function either with lifestyle modification or medications. To deliver robust 21st century diabetes care, it is important to personalize the treatment approach from the bottom up, starting with the lifestyle and habits of the individual. This knowledge should inform the treatment approach rather than a "let's see what happens" method. As we come to understand diabetes with more precision, it will be necessary to use a personalized, lifestyle-oriented, data-driven method of care rather than a physician-centered model. This is similar to what is taking place with precision medicine for cancer therapy.

So…where do we go from here? To better treat diabetes, it is crucial to take a more thoughtful approach to managing it. That starts with a personalized, supportive approach to care that includes glucose monitoring at the patient level, which is the topic of the next chapter.

Chapter 3

Glucose Monitoring

A s noted in the previous chapter, glucose is a simple sugar that is a primarily an energy source in living organisms. It is also a component of many carbohydrates. It is simple in the notion that we can easily define it chemically, but it is an extremely complicated source of energy when interacting with your body. In fact, high glucose levels are sometimes unrecognized by individuals with diabetes, but can have consequences over years, leading to serious eye, kidney, and nerve disease. When you have diabetes, your body's ability to produce or respond to the hormone insulin is impaired, resulting in abnormal metabolism of carbohydrates and elevated levels of glucose in the blood and urine. Elevated glucose levels can cause a myriad of symptoms (e.g., excess urination, blurred vision, and numbness in the feet to name a few), and if high enough, can threaten your overall health and life.

Thus, for those with diabetes, it is crucial to monitor and maintain your glucose levels. Often, patients are tasked with checking their blood sugar twice daily, typically before breakfast and before dinner. This is somewhat useful for patients with good control of their diabetes, and worthless for patients with uncontrolled blood sugar. The reason why is that twice-daily blood sugar testing provides information before meals, the low points during the day. We lack information about how high your sugar goes after meals, which reflects the ability of the pancreas to compensate. When blood sugar control is good to excellent, obtaining twice-daily blood sugar readings is only wasteful, but not harmful. The healthcare provider does not use this information to make a treatment decision. However, when you have uncontrolled blood sugar, obtaining twice-daily blood sugar readings can mask hyperglycemia after meals; if you don't see it you won't know it. Therefore, we should take a different approach to glucose monitoring to prevent harm from unrecognized high blood sugar. Additionally, a minimalist approach to controlling blood sugar is important so the patient can reasonably accomplish the goals. For these reasons, blood sugar monitoring for all patients with diabetes can be simplified into two methods:

1. Trend Analysis: Obtaining a representative sample of blood sugar values before breakfast and two hours after each meal, two to three times per week.

2. Active Management: Obtaining seven to fifteen blood sugar readings daily to manage insulin dosing and carbohydrate intake or using a continuous glucose monitor for real-time blood sugar management.

Let's unpack each in greater detail.

Trend Analysis

Trend analysis is my favorite exercise in managing diabetes because it gives the patient control over the disease. A typical trend analysis of self-blood glucose monitoring (SBGM) is three days per week; four times daily with finger stick readings before breakfast plus two hours after each meal. This dataset becomes superimposed on the LIFEMAP, so all the information is in one place and available for review.

Patients write down each blood sugar reading on paper, in a logbook, or simply upload it to a cloud-based platform through a smartphone app (a cloud-based diabetes management platform is forthcoming). This allows the information to be readily interrogated at the next office visit or via a telemedicine interaction. Patients find this approach convenient and useful because they now can evaluate their blood sugar after eating food, when it tends to be highest.

In addition, this practice avoids "data overload," which leads to confusion and difficulty interpreting their data. It is important to reemphasize that trend analysis is not a random construct, but is rooted in the insulin profile shown in figure 1. Through trend analysis, we are taking a small snapshot of beta cell function and its consequences in the post-prandial state with or without medication. If there is a deficiency in beta cell function, then there will be an elevation in post meal blood sugar. Medical providers can easily correct this with a treatment approach focused on one or several meals. Treatment success can then be quantified by observing the decrease in post meal glucose levels over time. Fasting blood sugar allows us to evaluate hepatic glucose production overnight and informs the healthcare provider about insulin dosing at bedtime (or other medications) that form part of their treatment plan.

For patients on oral medications with well-controlled blood sugar, trend analysis allows them to maintain a view of their glucose dynamics on a weekly basis so they maintain good glycemic control over months and years. At times, I negotiate with my patients that they can measure glucose two days per week; four times daily or one day per week; four times daily if we have confidence that glycemic control is well maintained. Here, the goal is to do the least amount of work necessary to get the best outcome.

Additional benefits to trend analysis include:

1. Structured approach to testing blood sugar at relevant times.
2. Personalization of which days and times to test blood sugar.
3. Minimizing the amount of testing.
4. Providing robust information to the healthcare provider that goes directly into the treatment plan.

Trend analysis is a useful tool because it requires ~12 glucose test strips per week (four strips per day x three days per week), which is possible when the patient receives 14 to 21 strips per week (two strips per day x seven days per week or three strips per day x seven days per week). Trend analysis requires a repeat office visit in three to four weeks or a smartphone interaction (in the near future) during this time period to review the blood sugar results. Adjustments to insulin dosing or oral medications can be made at that time depending on the pre- and post-meal glucose values.

For example, if fasting glucose is consistently high, then the patient requires additional insulin at bedtime or can be started on an insulin shot at bedtime. If post meal glucose is high, then it is worthwhile to pick one meal and change one insulin dose before that meal to get that post meal glucose under control. This stepwise approach to excellent blood sugar control allows the patient time to integrate new treatments into her lifestyle and to see benefit in real time as she continues to do her trend analysis.

Moreover, for patients who are newly started on fixed-dose insulin strategies, trend analysis works incredibly well for the majority of patients. This is because the blood sugar readings after meals informs the provider about whether too much or too little insulin is prescribed before the meal, while fasting blood sugar allows the provider to understand whether bedtime insulin dosing is correct. It is important to make the patient aware that a fixed-dose insulin regimen is not personalized treatment, but rather a general estimate of insulin requirement based on the healthcare provider's experience or a weight-based approximation; a physician-centered approach. Often it is the case that insurance plans will allow one test strip per day for patients not using insulin and three strips per day for patients on insulin. This severely limits the ability to check blood sugar more frequently.

Using this approach one can creep toward excellent glycemic control and potentially mitigate hypoglycemia. It is also worthwhile considering short-acting insulin secretagogues (oral medications that help boost pancreatic insulin secretion) or pre-meal saccharidase inhibitors (oral medications that block carbohydrate absorption in the intestine) in attempting to control post meal glucose for those individuals not on insulin.

But trend analysis is not without limitations. There is a conundrum for patients who take fixed-dose insulin; a lack of pre-meal glucose before lunch and dinner leads to guessing the insulin dose before the meal. There is risk of low blood sugar in taking a fixed dose of insulin before each meal without a blood sugar reading prior to that meal. If the patient has poorly controlled hyperglycemia, then most of the risk is mitigated because the likelihood of hypoglycemia is small.

Another pitfall of trend analysis is that patients often provide incomplete data. In this case it is important to reinforce the need for home glucose monitoring to ensure that overall glucose control is maintained. A special consideration is also important for a variety of patients who have conditions that alter red blood cell turnover, as this will affect hemoglobin A1C readings (the standard way of assessing blood sugar control over three months). For examples, patients with anemia, patients on kidney dialysis, organ transplantation, or red blood cell disorders all have falsely low A1C levels. This leaves SBGM as the only legitimate tool to assess overall glucose control. Assays like fructosamine and other enzymatic tests give short-term indication of glucose control and do not replace home SBGM.

In certain situations, where trend analysis identifies a problem with the blood sugar after dinner, the trend analysis can be switched to fasting blood sugar, two hours post breakfast, before dinner, and two hours post dinner. In this case, where post dinner hyperglycemia was identified with conventional trend analysis, switching the blood glucose monitoring to before dinner and two hours after dinner allows the clinician to build a bolus insulin scale before dinner to control post meal glucose. This approach is very effective and results in excellent control after dinner and in the morning. An example of this change in trend analysis is shown in case 6 later in chapter 12.

Trend analysis can also play an important role with those patients on dialysis. Patients with diabetes on hemodialysis require special considerations. Typically, dialysis occurs on Monday, Wednesday, and Friday, or Tuesday, Thursday, and Saturday, and patients are usually asked to be at the dialysis center by 7:00 or 7:30 a.m. Therefore, they require two LIFEMAPS; one for dialysis days and one of non-dialysis days. This often leads to two different insulin protocols if they are on insulin injections or an insulin pump. There are also special considerations for treating insulin-dependent patients on dialysis because they often demonstrate significantly lower glucose levels during dialysis than over the same time period on non-dialysis days.[23]

Moreover, dialysis patients often require lower insulin dosing because they lack renal clearance of insulin which accounts for ~50% of the removal of insulin from the circulation.[24]

For patients on oral medication, trend analysis becomes a simple and convenient way to assess their overall glucose control over months and years. Patients will get a better snapshot to consider if their medications are maintaining good glucose control through fasting and postprandial glucose measurements. It is simple and can be negotiated to two days per week;

23 M. Joubert, C. Fourmy, P. Henri, M. Ficheux, T. Lobbedez, and Y. Reznik, "Effectiveness of continuous glucose monitoring in dialysis patients with diabetes: the DIALDIAB pilot study," *Diabetes Research & Clinical Practice*, 2015, 107(3):348-354.

24 W. C. Duckworth, R. G. Bennett, and F. G. Hamel, "Insulin degradation: progress and potential," *Endocrine Reviews*, 1998, 19(5):608-624.

four times daily or even in the extreme one day per week; four times daily if the doctor and patient have full confidence that glucose control is well established over time. An example of Trend Analysis is shown in figure 2.

Case for Self Blood Glucose Monitoring

Trend Analysis
3 times per week/4
times per day

Glycemic Management

Glucose Profiling

Active Management

Active Management is the second choice when considering uncontrolled blood sugar management and reflects the other side of the tee-ter-totter shown in figure 3. Active management of diabetes, either type 1 diabetes or insulin-requiring type 2 diabetes, implies that certain pieces of information are essential for controlling blood sugar. As a standard, the bar is set at 7 blood sugar tests per day because information is needed before and after each meal and at bedtime. When thinking about adjusting insulin dosing, there are two factors that are buried in the dose:

1. The ambient glucose level and
2. The carbohydrate load of each meal

The ambient glucose level implies that a patient measures blood sugar before eating food. The carbohydrate load or glycemic index of food allows the diabetic patient to "guesstimate" or precisely measure the effect of the carbohydrate load on post meal glucose. Insulin pump manufacturers have developed algorithms to estimate insulin bolus dosing by inputting ambient glucose level, carbohydrate to insulin ratio, and sensitivity index. These parameters can be changed and are only useful if the patient understands how to reprogram the pump settings. Most patients have limited understanding of how the carbohydrate-to-insulin ratio works or how the sensitivity index works and therefore they do not adjust their pump setting unless it is with the healthcare provider.

For their part, nutritionists, certified diabetes educators, and diabetologists ask patients to count their carbohydrates before meals and adjust insulin dosing for this number. In a common scenario the healthcare provider and compliant patient devise a carbohydrate-to-insulin ratio and use that to calculate a pre-meal insulin dose.

Although guidelines from most diabetes authorities still use carbohydrate counting, this approach fails in the many diabetic patients and becomes a burden for most because of the demand to know the carbohydrate content of every food source they eat.[25] New recommendations are gradually creeping into diabetic practice where the patient is asked to estimate the carbohydrate load of each meal as low, intermediate, or high. This is an easy exercise most patients can master. With this information, patients can adjust insulin dosing up or down by 2-10, units depending on their food sources. Patients with significant insulin resistance who use high-dose insulin must make bigger adjustments on occasions because their insulin-responsive tissues (e.g., liver and skeletal muscle) are more resistant to insulin action.

Patients find this approach user friendly and can usually guess their carbohydrate load before a meal. In reality, few people take the time and effort to count carbohydrates or to learn this approach. For those who

25 K. J. Bell, A. W. Barclay, P. Petocz, S. Colagiuri, and J. C. Brand-Miller, "Efficacy of carbohydrate counting in type 1 diabetes: a systematic review and meta-analysis," *Lancet Diabetes and Endocrinology*, 2014, 2(2):133-140.

learn carbohydrate counting, it is my experience that they guesstimate their carbohydrate load before meals most of the time. As an obvious note, a post meal blood sugar allows the patient to see how they did in estimating their pre-meal insulin-dosing requirement.

Active management often reveals useful patterns of hyper- and/or hypoglycemia that can be acted upon by patient or healthcare provider. For example, a consistent pattern of hyperglycemia after dinner implies under dosing at the dinner meal, overeating carbohydrates, or significant insulin resistance. But with active management, the patient and physician now have a focus for diabetes management to interrogate the dinner meal and mange blood glucose into a better post meal range. For most patient with type 1 diabetes, active management is essential to achieve excellent glycemic control. However, Medicare and Medicaid allow only three tests strips per day in most cases for insulin-treated patients and one test strip per day for individuals who do not use insulin (CMS, 2018).

In addition to blood sugar testing before and after meals, a blood sugar test is needed at bedtime to unravel a "misconception" about long-acting insulin that is perpetrated by most pharmaceutical companies that make such insulin.

Case for Self Blood Glucose Monitoring

Glucose Profiling

Active Management
3-7 times daily

Glycemic Management

The Myth of Bedtime Insulin

In some ways it is both remarkable and mundane that patients have figured this out, but physicians and other healthcare providers have not. Fixed-dose bedtime insulin often does not work! We see the common practice of treating physicians providing patients with a prescription for 15 units (for example) of a long-acting insulin at bedtime, around 10-11 p.m. The patient begins to recognize that they develop hypoglycemia at 3:00 a.m., when their bedtime or post dinner glucose level is low normal 70-100 mg/dl. More often, the patient does not measure bedtime glucose, so they lack information that would otherwise inform the healthcare provider and patient that they are overdosed with basal insulin at bedtime. The patient does one of two things after two or three hypoglycemic episodes:

1. Stop the bedtime insulin or
2. Decrease the dose of bedtime insulin

Patients are often afraid to adjust their insulin dosing and are not forthcoming with their healthcare provider about their tweaks unless directly asked. For my patients, I have recently implemented a loose basal scale for bedtime insulin that looks something like this:

Blood glucose (mg/dl)	Insulin dosing at bedtime
100-150	10 units
151-200	15 units
201-250	20 units
251-300	30 units

The dosing schema can be personalized since hepatic glucose production and insulin resistance will vary for each individual. Clearly, insulin resistance and other factors impact the dosing regimen, but it has become quite clear to me that a fixed dose of insulin at bedtime is often incorrect. Patients are thrilled to adopt this method of administering basal insulin with a bedtime scale because they have validation of fear they have lived with for a long time unbeknownst to the healthcare provider and now their suspicion is addressed. As hypoglycemia is the most-feared complication of insulin treatment occurring in 30-80% of patients depending on the underlying circumstances,[26] patients are glad to have this potential risk somewhat mitigated. We might not readily admit it, but sometimes the patient is smarter than the healthcare provider.

To have any fighting chance of controlling diabetes, you have to first understand the relationship between your diagnosis and your glucose levels. Your body reacts to the rise and fall of glucose levels, and attempts to compensate for the shifts. This can create a myriad of symptoms, and even death if left completely unmanaged.

Thus, the two methods for considering uncontrolled blood sugar management, trend analysis and active management, can position patients to truly understand how to monitor and eventually control glucose levels. In doing so, they can avoid many of the dire consequences of rapidly rising and falling blood sugar levels. In resolving this issue, we can then focus on the topic of the next chapter, problems with hemoglobin A1C testing.

---

26   N. N. Zammitt and B. M. Frier, "Hypoglycemia in type 2 diabetes," *Diabetes Care*, 2005, 28(12):2948-2961.

Chapter 4

Problems with Hemoglobin A1C (A1C) Monitoring: When It Doesn't Work

I n the previous chapter, we discussed self-blood glucose monitoring as a critical activity for maintaining glucose control. Adding to this conversation, when your doctor draws blood in a commercial laboratory setting, they will test hemoglobin A1C (A1C) levels of those with diabetes. Hemoglobin A1C monitoring has been around for approximately 35 years and is a foundation stone of modern diabetes care. A1C reflects the amount of sugar that attaches to the outer surface of circulating red blood cells over a three-month time period.

When measured every three months, A1C reflects the endogenous glycosylation of red blood cell proteins that correlates directly with overall blood sugar control. A1C is elevated when the blood sugar over a three-month average is high. A1C will be low when the overall glucose control is good over a period of three months. The scale for A1C ranges from approximately 5% to 20%. I use the following rubric for A1C with my patients: 6.0% to 6.9% excellent, 7.0% to 7.9% good, 8.0% to 8.9% fair, 9.0% to 9.9% poor, and ≥10.0% uncontrolled. A1C has become a benchmark for modern-day glucose management because it is used for all outcome studies linking diabetes control to small blood vessel (eyes, kidneys, and nerves) and large blood vessel (heart attacks, strokes, and peripheral vascular disease) complications.

For example, in the classic Diabetes Control & Complications Trial (DCCT) of type 1 diabetes, approximate 1440 study subjects were randomized to a convention treatment group (typically two shots of insulin daily) or an intensive treatment group (typically 3-4 shots of insulin daily or an insulin pump) and followed for 6.5 years. The average A1C in the standard treatment group was ~9% while the average A1C in the intensive group was ~7%. At the end of the study there was a 76% reduction in diabetic eye disease,[27] a microvascular (small blood vessel) complication in addition to reduction in other microvascular endpoints. This study solidified the use of A1C as a benchmark of diabetes control.

Unfortunately, hemoglobin A1C has significant limitations because it relies on average red blood cell turnover of approximately 120 days, which, as stated, is the average lifespan of a red blood cell in the circulation. Many disease states that are comorbidities of diabetes can alter the red blood cell turnover rate (usually increasing the turnover rate) and as consequence, falsely lowering A1C. This can lead to under treatment of

27 Diabetes Control and Complications Trial Research Group, D. M. Nathan, S. Genuth, J. Lachin, P. Cleary, O. Crofford, M. Davis, L. Rand, and C. Siebert, "The effect of intensive treatment of diabetes on the development and progression of long-term complications in insulin dependent diabetes mellitus," *New England Journal of Medicine*, 1993, 329:977-986.

diabetes because the healthcare provider erroneously believes the glucose control is better than it really is. This leads not only to unrecognized poor blood sugar control and the potential consequences discussed above, but also the patient is told the blood sugar control is good to excellent.

Several common diseases that can falsely lower A1C include anemia, chronic kidney disease, blood transfusion, congenital disorders of hemoglobin production, and liver disease. Iron deficiency anemia is the most common form of anemia worldwide. It is often seen in menstruating women. Studies looking at both men and women with iron-deficiency anemia and A1C levels are mixed; some suggest that A1C is increased while other studies show decreased A1C levels.

A mechanism for these divergent study results has not yet been identified, but it is nevertheless important to confirm that A1C is an accurate reflection of overall blood sugar control in individuals with diabetes and iron deficiency anemia. Chronic kidney disease is also very common in patients with diabetes. Chronic kidney disease shortens the lifespan of circulating red blood cells and also decreases bone marrow production of non-glycosylated new red blood cells called reticulocytes. A1C cannot be used to accurately assess glucose control in individuals with diabetes and chronic kidney disease. In absence of A1C measurement, what can be done to assess the overall blood sugar control of an individual with diabetes?

The answer to this question comes from the LIFEMAP. Indeed, other assays for estimating overall blood sugar control are not satisfactory, as we have discussed. The LIFEMAP enables the healthcare provider to obtain a quality sampling of home blood sugar measurements (I refer to blood sugar measurements four times in one day before breakfast and ~2 hours after each meal as a "dataset") that can then be converted to an average glucose and then an average estimated A1C using the formula: A1C = (Avg Glucose + 46.7) ÷ 28.7.[28] The utility of having average glucose or

---

28 D. M. Nathan, J. Kuenen, R. Borg, H. Zheng, D. Schoenfield, R. J.
    Heine; the A1C-Derived Average Glucose (ADAG) Study Group, "Trans-

estimated A1C cannot be overstated because short of this information there is no other way to assess the month-over-month glucose control for an individual with diabetes. If you have a sampling of the lowest blood sugar during the day (potentially the fasting blood sugar) and the high blood sugars during the day (1-2 hours after meals), then you have all you need to estimate overall blood sugar control.

The dataset as constructed in the LIFEMAP is quite simple.

If we measure blood sugar at the four critical times of the day (before breakfast and two hours after each meal), then we indicate these numbers represent blood sugar readings on an average day in the life of an individual with diabetes. If we obtain a dataset two or three days per week, then we are saying that these days are representative of other days on average when we do not measure blood sugar.

Therefore, we have created a "trend analysis" by obtaining SBGM three days per week; four times daily; twelve blood sugar readings per week. These numbers can be averaged to obtain an average glucose (AG), and this number can be compared every month to see if AG is getting better or worse.

We can also convert AG into an estimated A1C value (using the formula shown above) which can then be assessed every 3 months when a routine hemoglobin A1C level is unreliable.

Therefore, the LIFEMAP offers us a path towards excellent glucose control when other conventional methods fail.

---

lating the A1C assay into estimated average glucose values," *Diabetes Care*, 2008, 31(8):1473-1478.

Chapter 5

Practical Application of Diabetes Care: 10,000 Feet or Ground Level

I n the previous chapter, we discussed one critical aspect of diabetes care concerning hemoglobin A1C. We learned about important instances where A1C cannot be used to assess three-month glucose control. But the good news is that the LIFEMAP can help to overcome this limitation by providing an alternative way to assess long-term diabetes control. The LIFEMAP provides other important aspects of diabetes care that enable individuals to maintain control over their sugar instead of their sugar controlling them. This means we can achieve better diabetes care by using a personalized approach to care that takes into account the lifestyle, habits, schedule, limitations, food choices, and feelings of an individual.

Later, we will discuss the impact of social determinants of health (e.g., money problems, environmental problems, alcohol and drug abuse, mental illness, anxiety, etc.) on diabetes control and compliance. I refer to it as "ground level" care when taking about a personalized approach to diabetes. Alternatively, 10,000-foot care is the impersonal, physician-focused, drug-centered diabetes care because it looks down from above and is quick to administer a solution without individualized information about the patient.

The decision to personalize diabetes care involves a balance in perspective between a global 10,000-foot view to ground-level considerations. In some instances, where diabetes is easily controlled with one or two medications, it might appear that 10,000-foot care is the better choice because it does not take a lot of time for the healthcare provider and the patient gets results.

However, diabetes is not a static, binary disease. For most individuals, diabetes is described as a disease that you either have or don't have, a binary disease. Unfortunately, diabetes has a wide spectrum of severity ranging from mild, easy to control; to severe, uncontrolled. All individuals with diabetes are on a spectrum of disease with a trajectory that can get worse over time. Unless you have a treatment approach that tracks the progression of diabetes or lack thereof you are not getting robust care. This is exactly what the LIFEMAP does; it provides a treatment approach and a tracking approach so that easy-to-control diabetes does not become uncontrolled diabetes.

Ultimately, most patients benefit from treating physicians and healthcare providers sculpting a treatment plan that considers both the big picture issues, as well as the more specific concerns most patients will almost certainly encounter. This chapter will explore in greater detail and demonstrate exactly why the intersection of the 10,000-foot view and the ground-level perspective can greatly benefit the overall health and treatment protocol of all patients afflicted by diabetes.

When looking at patients from the 10,000-foot view, I mean that for a particular patient who comes to my practice with an A1C of 9.5% (in my rubric 9.5% means poor diabetes control), we know that certain diabetes medications can lower A1C by an approximate percentage, usually 0.5% to 1.5%, thereby leading to better blood sugar control. We also know that diet and exercise are critical components of diabetes management and can impact significantly on drug efficacy. Therefore, a common 10,000-foot approach is to give the patient a glucose-lowering drug (often dependent on what their insurance covers), and then recommend that they reduce their carbohydrate consumption and exercise several times weekly. It is often an arbitrary practice to choose from a host of therapeutic possibilities, but there are some common practices as well. For example, metformin is probably the first-line drug of choice for mild to moderate type 2 diabetes (A1C levels between ~6.5% to 8.0%), especially given its ease of use, relative positive efficacy, low cost, and tolerable side effect profile for most patients.

Another common practice from the 10,000-foot view approach is to see the patient as an admixture of risks that can be mitigated by statins (cholesterol-lowering drugs), ACE/ARBs (kidney-protective drugs), and glucose-lowering agents, since these factors ultimately determine bad disease outcomes like eye disease, kidney disease, nerve disease, heart attacks, strokes, and vascular disease. At this level of care, physicians do not need an in-depth knowledge of the life schedule of the patient, which would offer a more personalized view. Rather, doctors rely on evidence-based medicine that lists the standards of care for diabetic patients. Many oral therapies potentially fit into this group of glucose-lowering drugs including DPP-4 inhibitors (drugs that help the pancreas produce more insulin), sulfonylureas (older, long-acting drugs, that directly stimulate insulin production from the pancreas), insulin secretagogues (newer short-acting drugs that directly stimulate insulin production from the pancreas), saccharidase inhibitors (inhibitors of enzymes in the bowel that help digest carbohydrates), incretin hormones (injectable agents that indirectly stimulate insulin production from the pancreas, slows the digestion of food in the stomach, and in some cases gives an "I'm not hungry" signal in the brain), SGLT-2 inhibitors (oral drugs that cause glucose to be disposed of in the urine), PPAR-γ agonists (agents that work to lower insulin resistance) and bromocriptine analogues (oral drugs that work on the brain to indirectly improve blood sugar levels).

Medical economics and physician preference, rather than the pathophysiology of disease or a personalized view of the patient, usually drive the decision points for the appropriate therapy to apply. This approach works well for many individuals in the short term. However, it is absolutely not the best practice for dealing with a chronic illness like diabetes because it lacks an active method to track the disease trajectory. I often see patients who had excellent A1C levels for several years who were followed every 3-4 months by their care provider. Over time, the A1C rises, but modestly until the patient has another medical illness and all of a sudden, the blood sugar is out of control. This process appears to happen suddenly, and sometimes it does. More often, the gradual rise in blood sugar over months to years that was not appreciated beforehand led to uncontrolled blood sugar. It would be nice to have a way to monitor your blood sugar trajectory more closely with less effort so that these unexpected events do not happen. The LIFEMAP does this. However, I have never seen any patient with a LIFEMAP other than my diabetic patients.

In my opinion, the better practice is to build a sustainable roadmap for the patient that provides various contingencies for safety, efficacy, and personalized choice. It is also possible (although not proven yet) that a more personalized view of diabetes care will lead to better outcomes and a prolonged insulin-free period. We learn from the UKPDS study that ~50% of diabetic patients on oral medication will transition to insulin over a 6-year time period.[29] Recently, SGLT-2 inhibitors and GLP-1 agonists have demonstrated beneficial cardiovascular effects, giving additional benefit to treating diabetic patients from a more global 10,000-foot view. Thus, a 10,000-foot approach to diabetes treatment can be practical, beneficial to the patient with cardiovascular risk, and often efficacious. However, without a ground-level approach that monitors the trajectory of diabetes, the 10,000-foot approach to diabetes care always falls short.

As it turns out diabetic medication can be structured according to the details obtained from the ground-level view. As a matter of practice, most physicians and healthcare providers use a mixture of these two approaches. For example, a patient with gastroesophageal reflux disease (acid indigestion) might be a poor candidate for metformin because it can provoke upper gastrointestinal (GI) symptoms. Therefore, a sulfonylurea, DPP IV inhibitor, insulin, or insulin secretagogue should be considered. GLP-1 agonists might exacerbate GI side effects since this is part of their tolerability profile. Part of their mechanism of action is to slow the motility of the stomach. When making a drug choice for a patient, doctors should consider factors like tolerability or susceptibility to the side effects of particular drugs. For example, recurrent urinary tract infections or vaginal yeast infections in women would preclude the use of SGLT-2 inhibitors.

In addition to this 10,000-foot approach, healthcare providers can work to superimpose the notion of virtuous and vicious drug candidates (table 1) since vicious drugs tend to cause weight gain while virtuous drugs tend to facilitate weight loss. We can classify diabetes medications by whether routine use leads to weight gain (vicious) or weight neutral/

---

29   UK Prospective Diabetes Study (UKPDS) Group, "Intensive blood-glucose control with sulfonylureas or insulin compared with conventional treatment and risk of complications in patient with type 2 diabetes (UKPDS 33)." *The Lancet*, 1998, 352(9131):837-853.

weight loss (virtuous). The newer classes of diabetes drugs attempt to play on the side of weight neutral or weight loss, which has been a major success for diabetes therapies. Insulin should generally be used at the onset for someone with hemoglobin A1C above 10%, but causes weight gain in most individuals because it leads to fuel storage (blood sugar is stored in tissues instead of circulating in the blood). These are reasonable 10,000-foot decisions for healthcare providers and patients that will yield predictable results in the short term.

Virtuous and Vicious Type 2 Diabetes Drugs

Virtuous (Weight Neutral/Negative)	Vicious (Weight Positive)
• Metformin - • DPP-4 +/- • GLP-1 analog high dose -- • SGLT-2 -- • α-glucosidase inhibitor +/- • Bromocriptine analog +/- • Phentermine + Topiramate -- • Lorcaserin - • Bariatric surgery ----	• Sulfonylurea ++ • Insulin +++ • PPAR-γ +++ • Insulin Secretagogue +

Please note + and – are indicative of the relative effect on weight for each agent
(eg +/- is neutral; - implies a negative effect on weight gain)

From a practical perspective, it is much easier to treat a patient at 10,000 feet. It takes a great deal of time to develop a personalized profile for each patient. In certain instances, it is appropriate and perfectly fine to treat a patient from 10,000 feet. In fact, one might argue that each and every individual with prediabetes, polycystic ovarian syndrome (a common cause of menstrual irregularity in young women), or insulin resistance should receive an insulin sensitizer like metformin or pioglitazone (PPAR-γ agonist) or intensive lifestyle modification using the

10,000-foot approach. In fact, we have learned from studies like DPPOS and UKPDS that type 2 diabetes is a progressive disease and that behavioral modification can only get you so far. At or about two years of implementing lifestyle modification, patients often tire of following the same strict regimen of diet and exercise.[30] This is not meant to take away from the efforts of those individuals who are resilient and disciplined to maintain intensive lifestyle modification, but only to respect the reality that most people cannot maintain such vigilance. So in the end, most patients succumb to the same common challenges, indicating that a more personalized approach is often the best one.

Making decisions from 10,000 feet is how we have practiced diabetes care for the past 90 years, especially in the past when there were limited pharmacological choices for diabetes treatment. We were then further removed from the molecular age of medicine, which began in the 1980s with the discovery of gene cloning. Under the older paradigm, there was a healthy debate about questions like whether sulfonylurea (a drug class which is widely used to date and stimulates insulin secretion from the pancreas) was the best first line treatment choice, whether long-term use increased cardiovascular risk, whether metformin was safe to use in patients with impaired renal function, and whether insulin caused heart disease and/or cancer. We have partially resolved these issues over the past 10-15 years, while the pharmacological field has moved past these early drug standards. The small repertoire of drug options in the past made the personalization of diabetes care impossible, and allowed healthcare providers to "script" a limited number of therapeutic options. In addition, limited home data collection prevented healthcare providers from implementing a data-driven approach to diabetes care. However, that is no longer the case. The environment and culture has changed, and

---

30 Diabetes Prevention Program Research Group, "10-year follow-up of diabetes incidence and weight loss in the Diabetes Prevention Program Outcomes Study," *The Lancet*, 2009, 374(9702):1677-1686.

doctors do have the ability to offer a more ground-level approach to treat this challenging disease.

Ground Level

Data-driven treatment algorithms will inform medical decision-making in the 21st century. Some individuals have referred to this as personalized medicine. However, for diabetes management we are limited to certain basic tools like home glucose monitoring to control blood sugar. As seen in patients with insulin-dependent diabetes, insulin infusion pumps and continuous glucose monitoring devices are data-driven approaches to recapitulate a healthy functioning pancreas.

These advanced tools are rapidly transforming insulin-dependent diabetes into a data-driven (and manageable) disease that requires less guesswork and more computing power. Unfortunately, the cost, complexity, and lack of understanding of these tools limit their availability to the great majority of individuals with diabetes. Nevertheless, using the available data to inform decision-making in diabetes management is necessary and potentially valuable when configured in the proper framework for type 2 diabetes.

The ground level approach to diabetes requires an investment of time on the front end of care that pays dividends later on. Personalizing a diabetes LIFEMAP and developing a treatment plan takes some time and requires undoing many years of the "old-school" 10,000-foot approach. For example, many patients come into my office and we have the following dialog:

> Doctor: "Ms. Jones, how are you managing your blood sugar?"
>
> Ms. Jones: "I check my sugar twice a day, before breakfast and before dinner."
>
> Doctor: "What happens when you show your doctor your home blood sugar readings?"
>
> Ms. Jones: "She looks at the numbers."

Doctor: "And then what happens?"

Ms. Jones: "She gives me back my meter."

Doctor: "Ms. Jones, do you realize that you're wasting blood and information?"

Ms. Jones (with a shocked look on her face): "Doctor, I've been doing that for years."

Doctor: "Ms. Jones, are you or your doctor using the information to inform a treatment decision?"

Ms. Jones (with a puzzled look on her face): "I'm not sure what you mean."

Doctor: "What I mean is that all these years you've been diligently collecting all of this information from home, but the information isn't being used to directly manage your blood sugar."

Ms. Jones: "I understand, but this is what my doctor told me to do."

Doctor: "Yes, Ms. Jones. You've done a great job following instructions, but now I'm going to take a different path and build a strategy with you so we can use the numbers to make treatment decisions about how to manage your blood sugar and use the information to track your progress. You see, there are four critical time points in the day to measure your sugar, before breakfast and two hours after each meal. If these numbers are good, then your blood sugar control is good. And oh, by the way, you don't need to do this all days of the week because we want to do a trend analysis. This means that we sample random days during the week to approximate what happens on the day that you don't check your sugar."

Ms. Jones: "Why are there four special times during the day to check sugar?"

Doctor: "Fasting blood sugar is a way of seeing what your liver does overnight. Your liver is a sugar factory, and when you have diabetes the 'off' switch is broken. This is why you might have high blood sugar in the morning even when you haven't had anything to eat. The other three points are two hours after meals, which represent the high points for blood sugar during the day. If these four numbers are well controlled, then your diabetes is well controlled. I tell my patients to pick 2 to 3 days per week and check blood sugar 4 times daily according to the LIFEMAP that we built together."

Ms. Jones: "You know, doctor, I've had diabetes for 23 years and you're the first one who has ever spent the time with me to explain what I'm doing and why I should do it differently."

This is how a diabetes LIFEMAP conversation goes in my office with thousands of patients.

So to summarize, when I build a LIFEMAP with my patient I start with wake-up time, breakfast time, lunchtime, dinnertime, bedtime, and snacks (if relevant). Once these times are scripted, then I translate these times into "before breakfast" and "2 hours after main meals." These times become the time for blood sugar testing with an understanding that the times are an approximation of an average day. Precision is not necessary so testing blood sugar approximately 2 hours after a meal is good enough. It is not expected that blood sugar testing be done exactly 2 hours after a meal.

The next step is to designate specific days of the week for blood sugar testing four times. Again, the particular day of the week is not as important as consistency in performing routine self-blood glucose monitoring. Once the routine is established it should be maintained forever or until I sit with my patient and, together, we decide to change the LIFEMAP to improve the sugar control at one specific time point in the day. We will discuss in more detail later how and when to alter a LIFEMAP. Once the patient has adopted my strategy for SBGM, diabetes has been transformed from a physician-centered disease to a data-driven, personalized, patient-centered disease. Success is just around the corner now. More-over, a cloud-based LIFEMAP will soon be available that will make blood sugar data collection even easier and inform patients via smartphone texts of the agreed-upon times for testing. The blood sugar data gets stored in a patient and healthcare provider portal and average glucose and estimated A1C are automatically calculated. In the near future, patient and provider will have real-time access to their blood sugar control.

In the next chapter we will drill down into social determinants of diabetes and how "life" can get in the way of achieving optimal blood sugar control even when care has been personalized. We will also spend more time discussing how to build a LIFEMAP and how it can be easily configured for both patient and healthcare providers.

Chapter 6

LIFEMAP and Social Determinants of Diabetes: Ground-Level View

As we move through our detailed conversation on diabetes and the best treatment approaches, we now arrive at the LIFEMAP. The LIFEMAP is the cure for many of the challenges patients with diabetes face. It offers a roadmap to reinvent the way medical providers evaluate and treat those with diabetes and provides a long-term solution for sustaining excellent blood sugar control. This chapter will better explain the specifics of the LIFE-MAP, and demonstrate exactly how it will shift outdated and ineffective treatment strategies into a more patient-focused and tailored approach. No two people live the same life, neither is the way diabetes impacts the overall health of different individuals. Much of the negative impact of diabetes can be attributed not just to the physical ailment, but to the specific person fighting it. As you will see in this chapter, the way a patient with diabetes lives their life can greatly impact the seriousness of the disease. Moreover, social factors that sometimes cannot be controlled dramatically impact the trajectory of diabetes. To make matters even more challenging, few doctors account for different eating and sleeping patterns, life stresses, and social and behavioral problems, indicating a one-size-fits-all approach to diabetes treatment will never lead to high-quality, optimal care for their patients.

The transition from 10,000-foot care to ground-level care is obvious in some ways, and insidious otherwise. The LIFEMAP focuses on ground-level care, which becomes a part of a "personalized medicine" approach. This is a simple construct the healthcare provider builds, and it starts by asking the very important questions from the moment a diabetic patient starts his or her day. The LIFEMAP walks the patient through their breakfast, lunch, dinner, snacks, and bedtime, all while recording the approximate times for each of these events. In addition, the LIFEMAP tracks social determinants of disease (e.g., insomnia, substance abuse, mental illness, lack of money, lack of medications, lack of adequate housing, eating disorders) that can upend the most careful plan. One might ask why it is important to identify and record social determinants of disease if there is really nothing that can be done to fix these problems. I offer two possible reasons for recording these life-altering social determinants. At worst, the healthcare provider understands why the patient is not getting to their goals of care for diabetes. At best, the mapping of social determinants of disease provides an opportunity to change behaviors and hopefully one day reconfigure healthcare spending to treat the root causes of these problems.

It is often surprising that the LIFEMAP discussion provokes many important aspects of diabetes care, including shift working, insomnia, skipping meals, dialysis days, and erratic meal times, as well as unexpected schedules that people adopt in their lives. It is presumptuous to use terms like "breakfast, lunch, and dinner" and think that this implies to a meal set at 8:00 a.m., 12:00 p.m., and 6:00 p.m. In fact, breakfast might refer to a meal at 7:30 a.m., which could occur before the patient goes to bed after working a night shift from 8:00 p.m. to 6:00 a.m. It can be different for everyone, and that is precisely why a 10,000-foot view doesn't always best serve the patient. In this example, breakfast now becomes a distortion of the upside-down schedule a person adopts to provide the critical necessities to support his or her family. In doing so, using generalities can be detrimental to the patient's health.

Unfortunately, many shift workers develop metabolic syndrome, obesity, diabetes, hypertension, sleep apnea, and the like because of dysregulated circadian rhythms.[31] It becomes difficult to treat insulin resistance, and even more challenging to determine the correct medication dosage for each patient that lives a life filled with a non-uniform schedule. Equally important in this example is that shift workers often revert to a semi-normal schedule on the weekends, although they are impaired by a chronic sleep deprivation cycle. So, at times it is necessary to create two LIFEMAPS for shift workers—one for weekdays and one for the weekends. In any case, doing so can be a true difference maker in a patient's overall health and management of diabetes.

Similarly, a patient undergoing dialysis presents a particular challenge because their behaviors and patterns change from those days they

---

31 F. A. Scheer, M. F. Hilton, C. S. Mantzoros, and S. A. Shea, "Adverse metabolic and cardiovascular consequences of circadian misalignment," *Proceedings of the National Academy of Sciences of the United States of America*, 2009, 106 (11):4453-4458.

receive dialysis (often three times per week on Monday, Wednesday, and Friday) to those days they don't. Thus, they might need two LIFEMAPS to account for the substantially different patterns in each day. It is as if this group of individuals lives on two entirely different schedules, because dialysis typically starts early in the morning at 7:00 a.m. Therefore, the patient (with or without a significant other) needs to be up by 5:30 a.m. to arrive on time to the dialysis center. Breakfast is often a meal they eat at 6:00 a.m. before dialysis or later in the morning as a snack. On non-dialysis days, the patient and their significant other probably revert to a more normal schedule for their meals.

It is particularly important for healthcare providers to recognize these two different schedules when the patient is being treated with insulin (as are most dialysis patients). Dialysis patients are routinely treated with insulin alone because of the relative low side effect profile (short of causing low blood sugar). Moreover, by the time they have made it to the complication of end-stage kidney disease that leads to dialysis their pancreas no longer makes insulin. Therefore, it is imperative to develop two separate insulin schedules for a dialysis patient to optimize care and avoid treacherous hypoglycemia. There is an example of these two potential LIFEMAPS in the case study about Harry.

Once a treating physician builds the LIFEMAP, it is important to discuss self-blood glucose monitoring at home in order to obtain glucose readings at the optimal times. Treatment providers can implement a trend analysis approach for most patients who do not have type 1 diabetes. In other words, once we have the times stated above mapped out, then we can create a schedule for checking blood sugar that will inform the healthcare provider of general trends in the ups and downs of blood sugar. We can generalize these trends so that for most patients it is not necessary to check blood sugar every day. Therefore, we can use trend analysis to approximate what the blood sugar will look like on the "not checking days" from the pattern of blood sugars obtained on the "checking" days. We thereby construct useful datasets of blood sugar results from the LIFEMAP and then apply a treatment strategy for all days based on approximation. Remember, controlling blood sugar is not an all-or-nothing proposition, but rather a matter of gradual improvement week over week and month over month until the overall goal is achieved.

LIFEMAP and Social Determinants of Diabetes: Creating the LIFEMAP

The LIFEMAP is a simple tool to improve blood sugar control. It was intuitively derived only after studying many patients in my diabetes clinic who suffered from poorly controlled blood sugar, but did not have a roadmap for success. It was also the result of applying statistical data to blood sugar management. In a perfect world, patients would measure blood sugar at least seven times per day, before and after each meal and at bedtime. For more intensive management (typically with type 1 diabetes) they use a continuous glucose monitor that measures blood sugar every few minutes if necessary. In an imperfect world, blood sugar management should fit together with lifestyle, compliance, and cost. Moreover, Medicare allows for one strip test per day with patients on oral medications and three test strips per day for patients on insulin. These ridiculous constraints significantly impact the ability to actually achieve optimal blood sugar results.

Thus, we create a LIFEMAP for all new patients suffering from diabetes. The LIFEMAP is a simple construct that walks a patient through an average day, starting from the time they wake up and taking them through their meals, snacks, and until bedtime. A typical LIFEMAP can be found in figures 4 and 5. *(see next page)*

The LIFEMAP approximates an average day for each patient and thereby gives the provider a way to designate specific times for obtaining self-blood glucose monitoring at home. The provider can then integrate

LIFEMAP for Diabetes Patients

WU 7:00 am B 7:30 am L 12:30 pm D 6:30 pm BT 10:00 pm

SBGM 7:00 am 9:30 am 2:30 pm 8:30 pm

SBGM*		SBGM		SBGM		SBGM

Time 7:00 8:00 9:00 10:00 11:00 12:00 noon 1:00 2:00 3:00 4:00 5:00 6:00 7:00 8:00 9:00 10:00 11:00 12:00 mn

Breakfast   Snack ±   Lunch   Snack ±   Dinner   Bedtime

Social Determinants of Disease   Avg. Glucose   Est. A1C

☐ psychosocial ☐ financial ☐ access ☐ substance abuse ☐ compliance

Micro/Macrovascular Complications

☐ Neuropathy ☐ Retinopathy ☐ Microalbumin ☐ Macroalbumin

☐ CVD ☐ CVA ☐ PVD ☐ Carotid Disease

* Self blood glucose monitoring

LIFEMAP WITH SBGM*

Allergies none
Cigarettes none
ETOH none

BB 6:00 a.m.	AB 9:00 a.m.	BL	AL 3:00 p.m.	BD	AD 9:00 p.m.	BT
105	135		155		143	
93	118		152		168	
116	133		109		154	
88	122		148		150	
122	156		136		150	
115	110		128		165	

Wake up   Breakfast   Snack ±   Lunch   Snack ±   Dinner   Snack ±   Bedtime

Time 6:00 7:00 8:00 9:00 10:00 11:00 12:00 noon 1:00 2:00 3:00 4:00 5:00 6:00 7:00 8:00 9:00 10:00 11:00 12:00 mn

Social determinants of disease   Avg. Glucose 131.5 Est. A1C 6.2%

☐ psychosocial ☐ financial ☐ access ☐ substance abuse ☐ compliance

Micro/Macrovascular Complications

☐ Neuropathy ☐ Retinopathy ☐ Microalbumin ☐ Macroalbumin

☐ CVD ☐ CVA ☐ PVD ☐ Carotid Disease

* Self blood glucose monitoring

this data into the patient's LIFEMAP. Physiology dictates four critical time points to assess blood sugar:

1. Wake-up time to assess glucose production from the liver overnight.
2. Two hours after breakfast.
3. Two hours after lunch.
4. Two hours after dinner.

These then become the time points for SBGM. We will later discuss the superimposition of SBGM onto the LIFEMAP. The LIFEMAP provides a simplified roadmap to inform the provider about which medication is appropriate to treat the defect in a patient's blood glucose and then engages the patient in a user-friendly approach to glucose management that minimizes the data collection and informs the patient about how well he/she is doing. The LIFEMAP also provides the healthcare provider with an average glucose from the dataset obtained and an estimated hemoglobin A1C (estimated A1C). This information will be automatically provided in the future through a healthcare portal. It only takes ten to fifteen minutes to develop a LIFEMAP, or a collective 20-30 minutes to work through an introduction, brief history, prior medical record review if available, medication reconciliation, and LIFEMAP development. Think about that for a second. 20-30 minutes of a physician's time can sculpt an entirely unique and specialized treatment plan for a new patient fighting diabetes. As my future goal, we are building a cloud-based LIFEMAP to enable providers to treat diabetes at the top of their license and facilitate data flow from patient to the cloud to the healthcare provider and back and forth. The LIFEMAP will become a standard of diabetes care worldwide once we have built out the tools to make the process even easier for patient and provider. For now, the LIFEMAP can simply be scribed into the medical note by the healthcare provider.

We will return to the new patient visit in a separate chapter, but for now you should see just how obvious and easy it is to implant a LIFEMAP into your medical note. Remember, everything we do for a patient with diabetes is an attempt to recapitulate, as close as possible, the normal physiology of beta cell function.

Building a LIFEMAP

Building a LIFEMAP is quite easy and merges into the normal doctor-patient visit. We start building a LIFEMAP with a simple query as to when a patient gets up in the morning (on average), when they eat breakfast, lunch, dinner, snacks, and when they go to bed. Most people, being creatures of habit, follow similar patterns of eating and sleeping day to day so getting a ballpark routine for their daily schedule is useful.

It is often quite surprising to see the myriad answers we get to these simple questions, ranging from the shift worker who eats breakfast at 7:00 a.m. and then goes to sleep for four to six hours, to the "no eater" who snacks throughout the day and then eats a large dinner. Many patients skip meals, substitute snacks, or eat at unconventional times (e.g., dinner at 9:00 p.m. or 10:00 p.m.). Also, and not uncommon, many individuals with diabetes have severe insomnia and "graze" throughout the night. It becomes evident that using words like breakfast, lunch, and dinner have no meaning without a LIFEMAP to provide context.

The LIFEMAP can be used to approximate an average day in the life of the diabetic patient. Building a LIFEMAP becomes a critical foundation stone of the first diabetes visit. It allows the healthcare provider to perform "trend analysis" of ambient blood sugar at home throughout the day and before the patient returns for the second visit. It is also important to incorporate any behavioral influences in the LIFEMAP that might impact on diabetes or cardiovascular disease. These might include daily alcohol consumption, cigarette smoking, drug use, depression, food addiction, or other personal idiosyncrasies.

Once we complete the LIFEMAP, the healthcare provider can develop a plan for self-blood glucose monitoring. I like to use a trend analysis approach. It incorporates the LIFEMAP into the glucose monitoring and allows for an approximation for the day-to-day trends in glucose control. In general, I ask the patient to perform SBGM three days per week, four times daily, using the LIFEMAP to inform me of the critical times for glucose checking (fasting plus two hours after main meals). It is important to recognize that this is not an arbitrary construct, but is derived from the insulin secretory pattern of a healthy pancreas shown previously in figure 1. This approach gives my patients flexibility to choose which days they will check their blood sugar and allows them days off to recover from the constant pricking of their fingers. Many patients on oral agents can very effectively use this approach to maintain excellent blood sugar control once or twice weekly. I almost never prescribe twice-daily blood sugar monitoring for my patient because the normal beta cell physiology dictates the four-times-a-day approach.

Alternatively, patients who are on active management (meaning they need an ambient blood sugar measurement before meals to calculate a bolus insulin dose and after meals to see if their dose strategy was correct) require a minimum of seven blood sugar checks per day. This approach is common for patients on insulin pumps or multiple daily insulin injections. Here, we rely on blood sugar monitoring before each meal, two hours after each meal, before and after snacks, and then finally at bedtime.

The LIFEMAP is an integrative tool because it allows the healthcare provider to set the pattern of SBGM at home. It is important to emphasize that the LIFEMAP approach also recognizes and accounts for the limitations of glucose test strips that are provided to the patient by his/her insurer.

For example, engaging trend analysis, we devise a strategy for checking SBGM three days per week, four times daily in order to obtain useful information that can be acted upon on the follow-up visit (12 strips per week are well within the Medicare/Medicaid guidelines for three strips per day with insulin use).

A simple example of this integration goes as follows:

LIFEMAP

Wake Up (WU): 6:30 a.m.

Breakfast (B): 7:00 a.m.

Lunch (L): 12:30 p.m.

Dinner (D): 6:30 p.m.

Snake (SN): 8:30 p.m.

Bedtime (BT): 10:30 p.m.[32]

SBGM: 6:30 a.m., 9:00 a.m., 2:30 p.m., 8:30 p.m.[33]

The method we use is fasting blood glucose plus two-hour post meal glucose. The patient records these numbers at home, writes them down on a sheet of paper or logbook, and brings the results to their healthcare provider on the next visit. The trend analysis approach gives the provider integrated information about the four critical time points during the 24-hour period that goes into glucose control: fasting and post meal glucose.

However, there limitations of trend analysis. For example, a patient who is on a two-shot, three-shot, or four-shot insulin-dosing regimen needs to be on fixed doses of insulin before meals because they will be missing critical pre-meal data points that are needed to correct for the pre-meal ambient glucose level. In general, patients who have uncontrolled blood sugar benefit from this fixed-dose insulin approach as a first step towards better glucose control because they have a large buffer against hypoglycemia due to uncontrolled hyperglycemia.

As patterns emerge, the rubric of three days per week, four times daily can be transformed into a more focused approach. For example, a

---

32 WU = wake up, B = breakfast, L = lunch, D = dinner, SN = snack, BT = bedtime

33 BGM= self-blood glucose monitoring

SBGM logbook might show a pattern of blood sugars as follows:

BB	AB	AL	AD
148	220	256	188
168	205	195	238
136	188	265	310
162	234	258	296

Here the healthcare provider decides to focus on hyperglycemia after dinner. The patient is presently on a standard three-shot insulin regimen (example is show below):

1. Mixed insulin before breakfast (Humalog 75/25 or Novolog 70/30 given 5-10 minutes before breakfast; Humulin 70/30 or Novolin 70/30 given 40-60 minutes before breakfast).
2. Short-acting insulin before dinner (Novolog, Humalog, Admelog or Apidra given 5-10 minutes before dinner); Humulin R or Novolin R given 40-60 minutes before dinner, and
3. An intermediate- to long-acting insulin at bedtime (Lantus, Levemir, Tresiba, Humulin N or Novolin N).

Insurance constraints and drug benefits will often dictate which insulin the healthcare provider chooses.

The provider decides to build a bolus scale before dinner, but recognizes that a pre-dinner blood sugar will be necessary. With respect to the fact that trend analysis is an approximation and not precise, the provider changes the SBGM to before breakfast, two hours after breakfast, before dinner, and two hours after dinner, thereby giving her the necessary information to build a bolus scale before dinner on the following visit. The provider then builds a pre-dinner bolus scale that might look something like this (please recognize this is an arbitrary scale for illustration only and not to be used for your patients.) Physicians must personalize the scale to fit the patient's insulin requirements and it will be a "guesstimate" until the next visit when post dinner glucose data is presented:

Blood sugar	Insulin Bolus Dose
90-120	3 units
121-150	5 units
151-180	7 units
181-210	9 units
211-240	11-12 units
241-270	14-15 units
271-300	17-18 units
301-330	20-22 units
331-350	23-25 units

It is important to cross-check the insulin dosing with your patient to make sure he is comfortable with these doses. Otherwise you have a disconnect that leads to noncompliance or misunderstanding of the instructions. Patients will rarely confess to their healthcare provider that they are not following instructions. An open-ended approach that engages your patient will usually allow for freedom to use their own "good" judgment on insulin dosing.

For example, I always ask my patient if they are comfortable injecting 15 units of short-acting insulin before dinner when their blood sugar is 248 mg/dl. They will often say "yes," but sometimes negotiate the bolus dose with me so we get to an agreed-upon dose. This approximation can then be used as a starting point to build the pre-dinner bolus scale. On the next visit the new dataset allows the healthcare provider to adjust the insulin dosing so that excellent glycemic control is achieved. Since the bolus scale dose is an iterative process, I tend to under dose on the front end to avoid hypoglycemia and then gradually tweak the doses until adequate glucose control is achieved (e.g. ~120-160 mg/dl 2 hours after dinner).

When diabetes care is organized and patient centered using the LIFEMAP, patient compliance follows. My patients typically have two responses to the LIFEMAP on the follow-up visit: 1. "I have had diabetes for 20 years and nobody ever took the time to explain to me what I have to do to keep my sugar controlled." 2. "I check my blood sugar twice a day like I always do." For the first patient, they understand the instructions and made the necessary changes to obtain useful information that transforms their diabetes care forever. Once they have made the transition to the LIFEMAP method there is a dramatic and consistent improvement in their blood sugar control over the weeks and months. For my second patient, I reengage the patient and explain why I would prefer they check their blood sugar in a different way. We do a reboot and reinforce the concepts of the LIFEMAP. The lesson to be learned here is that long ingrained habits are hard to break. My second patient usually "gets it" the second time and adjusts to the LIFEMAP method of blood sugar checking. Once the transition is in place, then the fun begins of achieving excellent blood sugar control and transitioning the control from the healthcare provider to the patient. My patients are thrilled to be in control of their blood sugar and now possess the necessary tools to achieve success.

The LIFEMAP is the cure for the common (and uncommon) cases of diabetes. It offers patients a wonderful opportunity to gain valuable insight into how their lifestyle can impact their diabetes. As we monitor each specific case, we can learn a great deal of information and then adjust accordingly to ensure healthcare providers are properly managing the health of their patients from a very ground-level view. In doing so, we can then begin to adjust and discuss lifestyle choices, like how food choices and nutrition impact their diabetes as a whole.

Social Determinants of Disease

I used to call this term "behavioral drivers of disease" until a colleague informed me that many of the important issues that lead to poor care and poor health outcomes go way beyond behavior. For many today, issues of poverty, illiteracy, dangerous home environments, lack of access to care, and lack of quality food resources are all serious factors that can lead to bad outcomes when diabetes is superimposed. It would be naïve to believe that building a LIFEMAP for diabetes care can change these factors. In fact, many of these factors undermine efforts to improve diabetes care because the LIFEMAP is a low-priority item when considering that these other factors can be fatal. Nevertheless, building a treatment plan that includes social determinants of disease (SDD) can inform the healthcare team of why a particular patient is demonstrating noncompliance with medications or health instruction. It is my hope that by building a massive database of patients with clearly defined SDD we can eventually shift healthcare dollars from the back end of the process when disease is out of control and rampant to the front end of care where prevention and compliance is the goal. Obviously, for the healthcare system, cost saving will be a critical factor in shifting cost from the back end to the front end. Meanwhile, I will build my LIFEMAPs. We can now turn to a critical component of diabetes management, that of nutrition.

Chapter 7
Nutrition

To this point in the book, we have outlined the great challenges diabetes presents to those afflicted. Patient after patient struggles with the disease, and often attempts various natural cures and remedies while at the same time trying to live a healthy and happy life, even while fighting diabetes every day. There is a great deal of outdated and inaccurate data and research available, leading healthcare providers to feel as if treating diabetes is a one-size-fits-all approach. That is hardly the case.

Moreover, randomized clinical studies, considered to be the gold standard for diabetes care recommendations, are not personalized and reinforce a one-size-fits-all mentality. Randomized clinical trials provide important information yet have limitations in real-world diabetes care and can have bias when conducted by pharmaceutical companies to promote their own product.

There are a number of factors, both environmental and behavioral, that impact the manner in which physicians and healthcare providers should diagnose and treat diabetes. It is valuable to consider the global perspective of diabetes, which continues to increase year over year, but if we are to truly make great change, then we all have to look towards a much more ground-level approach. That is the ultimate goal and basis for the LIFEMAP.

The LIFEMAP is a powerful tool for healthcare providers and patients alike. It offers the best of both worlds, taking a global perspective that we know offers general guidelines (including randomized clinical drug trials), but then looking at the specific needs of the patient as well to supplement treatment. In this chapter, and those to come, we will build on the LIFEMAP and begin to supplement its strengths. So far, we have focused on managing blood sugar alone, but we know that blood sugar is affected negatively by stress, lack of sleep, sedentary lifestyle, and nutrition.

When it comes to diabetes, nutrition is an important aspect and vital to the overall health of the patient. How you eat is just as important as your nature or predispositions. The food you put in your body, or your nutrition, can play a pivotal role in regulating blood sugar and exacerbating your diabetic symptoms. From my perspective, nutrition is the road that led many to unintended diabetes and in part a road back. For the vast majority of individuals who develop diabetes, nutrition plays a key role in the disease process 95% of the time. It would be nice if we could turn back the clock and find a time when those extra five or ten pounds were just an annoyance.

Unfortunately, those pounds continue to increase as we age. Unless we put a preventive plan in place, unintended consequences eventually develop. I like to say the LIFEMAP is a plan intended to keep the "genie in the bottle." When a patient comes to me with prediabetes or diabetes during pregnancy, I am in full-tilt preventive mode. This is because healthcare professionals recognize that in time the genie escapes, diabetes ensues, and getting the genie back into the bottle is harder than keeping the bottle corked. Indeed, nutritional composition is linked with diabetes, obesity, cardiovascular disease, and possibly Alzheimer's disease, and by far, a Westernized diet is the worst. We also know that food composition is part of the solution to improved blood sugar control and decreased cardiovascular disease and that a Mediterranean-style diet is probably the best one (e.g., a diet rich in fish, nuts, extra virgin olive oil (uncooked), vegetables) when compared to a "low-fat" diet.[34, 35]

For whatever reason, many patients feel monitoring blood sugar is enough to regulate diabetes. But the truth is that diabetes is a disease fought meal by meal, as what you put into your body can tremendously impact how you feel, and ultimately how healthy you might be. Even more striking, a recent study demonstrated that food order can make a huge difference on blood sugar. Study investigators took a small group of individuals with type 2 diabetes and fed them the same diet with different food orders: carbohydrate first followed by protein and vegetables, protein and vegetables first followed by carbohydrates, or all foods at the same time. To their surprise, eating carbohydrates last (in this case orange juice and ciabatta bread) led to a dramatic reduction in blood sugar over the next three hours compared to eating carbohydrates first.[36]

34   J. Salas-Salvado, M. Bulló, N. Babio, et al., "Reduction in the Incidence of Type 2 Diabetes with the Mediterranean Diet: Results of the PRED-IMED-Reus nutrition intervention randomized trial," *Diabetes Care*, 2011, 34(1):14-19.

35   R. Estruch, E. Ros, J. Salas-Salvado, M. I. Covas, et al., "Primary Prevention of Cardiovascular Disease with a Mediterranean Diet Supplemented with Extra-Virgin Olive Oil or Nuts," *New England Journal of Medicine*, 2018, 378(25):e34(1-14).

36   A. P. Shukla, J. Andono, S. H. Touhamy, et al., "Carbohydrate-last meal pattern lowers postprandial glucose and insulin excursions in type 2 diabetes," *BMJ Open Diabetes Research & Care*, 2017, 5:e000440.

It is important to begin to integrate the nutritional assessment into the LIFEMAP so we can build a complete picture of how nutrition impacts blood sugar. I like to give an example that I see very often of patients who come from skilled nursing facilities. These individuals have very strict mealtime and snack schedules as well as wake-up and bedtime. What I eventually discovered is that all patients receive bedtime snacks! I ask my patients if they eat the snack because they are hungry or because it is offered to them. More than half say they are not hungry. So, I ask if they would mind if we stop the bedtime snack. Usually they agree to do so. Not surprising, the next morning their blood sugar dramatically improves. This is one example of why integrating nutritional information into the LIFEMAP is critical. If a dedicated nutritionist is available, then it is critical to obtain information that relates back to the overall treatment plan. This information might be the critical factor that makes the difference between blood sugar that is well controlled or uncontrolled.

At the first visit with a nutritionist, patients can acquire information regarding intake of sugar-containing beverages, outlining a meal plan for the three daily meals (or however many are eaten), and implementing preferred snacks into their eating patterns. These are all important because they can impact the rise and fall of your blood sugar. From the perspective of a physician, we prefer grouping meals into low, moderate, and high carbohydrate content for ease of patient compliance.

This is particularly important for patients who take insulin since they often adjust their pre-meal dose based on their expected carbohydrate intake. In my experience, most patients "eyeball" carbohydrate content of meals and adjust their insulin dose accordingly. Some patients perform a robust counting of carbohydrates, but they are still making an educated guess about the correct dose of insulin to inject because other factors influence the effect of insulin on blood sugar, including food composition and order as shown above, stress level, and absorption in the digestive tract. Regardless, it is important to consider carbohydrates because they are what raise blood sugar. In doing so, we can better regulate how diet and nutrition impact a patient with diabetes.

It is valuable to group food into a variety of categories to ensure we maintain healthy eating habits and moderate food consumption. Specifically, we group carbohydrate content as follows:

1. Low: <30 grams
2. Moderate: 31-75 grams, and
3. High: >75 grams

As a reminder, one gram of carbohydrate equals four calories, so 50 grams of carbohydrates is equal to 200 calories.

It is useful to give patients examples of what a typical low, moderate, and high carbohydrate meal might look like. Many of us like to learn by examples rather than generalities. For example, a simple menu plan that incorporates one or two examples of what three meals can look like can be easily modified to your personal taste and budget.

A low carbohydrate meal for breakfast is one scrambled egg, two slices of turkey bacon, one slice of plain toast (~15 grams for white bread and ~12 grams for wheat bread), and coffee with a nonnutritive sweetener.

A moderate carbohydrate meal includes a turkey or ham sandwich (~30 grams) with an apple (~10 grams), and a glass of milk (~15 grams).

A high carbohydrate meal includes spaghetti and meatballs or pizza.

These real-world examples can be "stretched" and modified to "guestimate" the carbohydrate content of most meals. To better illustrate this point, below you will find an example of a nutritional template that can be integrated into the LIFEMAP. This can often be done by a nutritionist or certified diabetes educator during your visit and integrated into the LIFEMAP.

Diabetes Nutrition Assessment

Information obtained from: patient
Food access/security: was on food stamps; recently discontinued
Cooking facility: adequate

Anthropometrics:
Weight: 289 lbs (131.3kg) Height: 5'7" (170 cm) BMI: 45.2 (morbid obesity)

Diet Recall:
Intake of sugar-sweetened beverages: 0 /day

Wake up: 10 a.m.
Breakfast: 2 hard-boiled eggs, cheese, apple or banana or pineapple , edamame, sometimes hash browns
B Carbohydrate content: 30-45 gm
Lunch: 1-2 p.m., tuna fish sandwich or turkey and cheese sandwich or 1 cup broccoli with cheese or salad + piece fruit
L Carbohydrate content: 45 gm
Dinner: 1-1/2 cups whole wheat pasta + ground beef + ricotta cheese + mozzarella + fruit + broccoli or peas or salad (mixed greens + croutons + cheese + spinach) + balsamic or blue cheese
D Carbohydrate content: 60 gm

Nutrition Diagnosis:
Problem: Impaired Nutrient Utilization; Related to: Diabetes
Evidenced by: Need for therapeutic diet, Hemoglobin A1C 10.3% (prior date) 7.9% (recent date)

Nutrition Intervention:
Recommended diet: 2000 calories. No concentrated sweets

- Identifying carbohydrate food sources
- Portion control

- Hypoglycemic treatment
- Continued cessation of sugar-containing beverages

Goals Established:
- Glycemic control
- Identify carbohydrate food sources + portion sizes

Total Time spent with patient: 25 minutes

Let's take a moment to point out some important parts of this nutrition assessment chart. First, you will notice that the healthcare provider should notice certain environmental factors, like considering the quality of the patient's cooking facility, as well as how they purchase their food. That is important to note, as it might be unrealistic to expect a patient on food stamps to regularly eat expensive produce and lean meats.

Next, outline the patient's anthropometrics, which includes details about their height, weight, and body mass index (BMI). These can be helpful measurable tools to outline the current state of their health.

Physicians or nutritionists can then plot sample eating habits in hopes of gaining a better understanding of exactly what the patient is putting in his or her body. This might be the most valuable piece of this chart, as you obtain a realistic glimpse into what they are eating and when they are eating it. This empowers you to offer direct feedback based on what's in front of you.

After doing so, you are then able to come to a general nutritional diagnosis, which can be based off the data in the chart. What do they eat? When do they eat it? How much does the patient weigh? These are all viable measurements that will help you formulate some important conclusions.

After doing so, the nutritional chart easily folds into the LIFEMAP, and the treating healthcare provider or nutritionist can then begin to sculpt an intervention program that assists the patient in eating in a way that supports health and minimizes the overall impact of diabetes.

And finally, the nutritional chart above sets specific goals for the patient and outlines what he or she can do to create and sculpt a better nutrition plan.

As you can see, in just 25 minutes a treating nutritionist or physician can secure a tremendous amount of information about the patient, eating habits, and overall nutritional health. The old saying is that which can be measured can be improved. In tweaking and utilizing this plan, we as treatment providers can better serve our patients by creating healthier eating habits to directly combat the ill effects of diabetes.

Much of the information outlined in this chapter is geared towards those patients living life outside of the hospital, and doing their best to fight diabetes. For most of them, they'd prefer to keep it that way. But not all patients can be so lucky. In the next chapter we will take a deep dive into the chronic care model, reserved for patients whose diabetes has progressed and created tremendous complications for their overall health.

Chapter 8

Chronic Care Model

With a better understanding as to how nutrition can impact a patient fighting diabetes, the next stop in our journey is gaining an understanding as to how the chronic care model impacts those same patients. A chronic condition is one that required ongoing adjustments by the affected person and interactions with doctors, other healthcare providers, and the overall healthcare system. The American College of Physicians indicates that the Chronic Care Model (CCM) is a multifaceted, evidence-based framework for enhancing care delivery by identifying essential components of the healthcare system that can be modified to support high-quality, patient-centered chronic disease management.[37] The CCM provides a

---

37 E. H. Wagner, "Chronic disease management: what will it take to improve care for chronic illness?" *Effective Clinical Practice*, 1998, 1(1):2-4.

systematic approach to practice transformation. Interrelated elements of the CCM include:

- **Health systems**, including culture, organizations, and mechanisms to promote safe, high-quality care.
- **Decision support** based on evidence and patients' preferences and needs.
- **Clinical information systems** to organize patient and population data.
- **Patient self-management support** to enable patients to manage their health and health care.
- **Community resources** to mobilize patient resources.
- **Delivery system design** for clinical care and self-management support, including team care.

While the CCM proposed by Wagner in 1998 is correct in many ways, it remains purely aspirational until we developed the LIFEMAP. Treating patients in the outpatient setting is vastly different than delivering acute care in the hospital. Therefore, the chronic care model requires a different approach entirely. However, the rubric of acute care continues to be carried over to the chronic care setting and leads to wasted time and ineffective (low-impact) care. When a healthcare provider is tasked with so many things to do for a patient it becomes impossible to do any of those things well. I divide the chronic care office visit components into three categories:

1. High-impact interventions
2. Moderate-impact interventions
3. Low-impact interventions

Although this creates a disruptive paradigm for healthcare providers, our present approach to chronic care delivery leads to low-quality outcomes. Why? The answer is simple—the model is wrong because it dilutes high-impact interventions with low-impact interventions. It is somewhat perverse that teaching the chronic care model (even in American medical schools today) is stuck in a rubric that was created over 100 years ago and includes steps like Chief Complaint, History of Present Illness, Family & Social History, Allergies, Medications, Review of Systems, Physical Exam, and Medical Decision Making (Assessment and Plan). While still the gold standard for assessing acutely ill patients in the hospital setting, this approach just doesn't work for chronic care like diabetes management.

The medical community and insurance payers have not fully contemplated that chronic care delivery, and diabetes care in particular, requires a different model than what is routinely used today. Treating diabetes is not as mathematical or systematic as this model would suppose it to be. As a doctor, it is important to have the opportunity to adjust and be flexible with the symptoms you might see. Depending on the seriousness of the condition, chronic care should be implemented at various levels. Not all patients need all elements of the acute care model in their chronic diabetes visit, yet we continue to teach this model as a one-size-fits-all approach to medical students. High-impact interventions should always lead in the chronic care model because they will yield important results for the patient if acted upon. Low-impact interventions should be let go, especially if their yield for improved quality of care is poor.

Obviously, the patient is in your office for some form of high-impact interventions. Thus, a healthcare provider should turn to a physical exam only for focused complaints on the first visit. Otherwise, this can be performed at a later visit or not at all, since it is frankly a low-yield intervention. As a reference point for discussion, here is an expansive list of impact interventions as they pertain to a diabetes visit:

High-Impact Interventions

In a setting where a patient needs high-impact intervention, the healthcare provider may consider the following steps:

- Obtain length of diabetes, microvascular complications, macro vascular complications, social determinants of disease, and comorbidities.

- Obtain vital signs and finger-stick glucose.
- Consider allergies, cigarettes (or other smoking habits), alcohol use.
- Medication reconciliation, including present treatment.
- A1C, lipids, urine microalbumin to creatinine ratio, if available.
- Build LIFEMAP and develop SBGM rubric for patient.
- Make sure patient has glucometer, test strips, and lancets.
- Give patient logbook with instructions (or in the future sign the patient up for the Concierge Care LIFEMAP).
- Attend to any emergent problems.
- Approximate time: 45-60 minutes (Coding level 4 or 5).

Moderate-Impact Interventions
- Disease-focused physical exam: eyes and feet if no complaints, possibly cardiovascular system).
- Manage comorbidities.
- Review of systems.

Low-Impact Interventions
- Routine physical exam.
- Family history (unless probing for genetic forms of diabetes).
- Other past medical history unrelated to diabetes care.

As healthcare providers, we lean on implementing a standard of care taught to us over many decades. This includes a rubric of history, physical examinations, and medical decision making. This rubric is embedded in the fabric of differential diagnosis and medical coding. It should be considered a foundation stone of robust patient care. However, recent technologies like Electronic Medical Records, Google Scholar, E-pocrates, Medscape, Up-to-Date, and other tools have changed medicine in subtle and explicit ways. Because of this trending technology, we have instantaneous information at our fingertips. When used properly, they can absolutely enhance patient care. Indeed, it is challenging to cover all the "necessary components" of a complete medical evaluation in a typical 45-60 minutes new-patient visit or a 15 to 20-minute follow-up visit. However, since these are now the benchmarks for chronic care delivery, we often have no choice. Therefore, it is necessary to consider new tools that prioritize the critical elements of diabetes if we are to deliver optimal care to our patients.

One point of discussion is the value of a heart and lung exam for a new patient with diabetes who provides no history of cardiovascular risk. This can often be a waste of time, mostly because it takes time away from critical elements of care. The heart and lung exam would be considered low-impact interventions and should not be included in a new-patient visit. Parsing the physical exam is important because there are only a rare few elements that actually provide value to the patient. These might include a foot exam and a fundus (eye) exam. Otherwise, elements that lack a chief complaint do not add to the overall care. Therefore, we should place the physical exam in in the "**low impact category.**" We will talk in more detail about these critical elements of physical exam and where they fit into my rubric of the LIFEMAP later in the book.

However, to demonstrate this point more concretely, I broke out my typical follow-up diabetes visit into its vital components. I then timed each aspect of the visit. Things like entry in the room and greeting the patient takes about **1-2** minutes, depending on the particulars.

I then review prior notes to make sure I covered all aspects of care that were not addressed on the previous visit.

One obvious point to reemphasize is that a specific complaint by the patient on the first visit should be attended to at that time. It is also a reasonable time to order lab tests, if not previously done, before the patient leaves the office so that baseline parameters like A1C, lipid profile, and urine microalbumin to creatinine ratio can be assessed. In addition, renal function and hepatic function need to be assessed in case the healthcare provider chooses to use an SGLT-2 inhibitor or statin, respectively.

In addition, for individuals with diabetes I find it important to review a self-blood glucose monitoring (SBGM) logbook or glucose meter from home if available. This activity typically takes **3-5** minutes.

Medication reconciliation is the next critical element of the visit. Even with my compulsion for a complete, "clean" medication record log, using electronic medical record leaves a lot of "litter" from other physicians and healthcare providers. By litter, I mean medications that were previously prescribed, refilled, dose adjusted, or prescribed, but not taken for a variety of reasons or not marked as treatment changed or treatment completed. These prior medications stay on the medication list in our electronic medical record and need to be cleaned up to ensure a treating physician can muddle through the information in front of him or her.

For example, frequently a patient is started on a statin drug and takes three to five doses before she experiences musculoskeletal pain and aching. She discontinues the medication, but neither the pharmacy nor the doctor is aware of this change in medication adherence. It is remarkable to understand that studies have demonstrated medication compliance rate of approximately 65-70% for patients with type 2 diabetes.[38] The statin drug stays on the medication list unless the healthcare provider actively removes it. Therefore, it is frequently the case that nobody knows which medications the patient is taking, including the patient and sometimes even the pharmacist. Medication error accounts for a part of "medical error." In the US alone, medical error has been estimated to cause ~200,000-400,000 death per year.[39] Thus, medication reconciliation becomes a **"high impact"** activity.

---

38   K. Iglay, S. E. Cartier, V. M. Rosen, V. Zarotsky, S. N. Rajpathak, L. Radican, and K. Tunceli, "Meta-analysis of studies examining medication adherence, persistence, and discontinuation of oral antihyperglycemic agents in type 2 diabetes," *Current Medial Research and Opinion*, 2015, 31(7):1283-1296.

39   M. A. Makary and M. Daniel, "Medical error—the third leading cause of death in the US," *British Medical Journal (BMJ)*, 2016, 353(2139):1-5.

As for the patient, it is typical for any diabetic patient to take between 10-20 pills ± or shots on a daily basis. When asked to review a computer-generated medication list, patients generally respond with something like: "I take the blue, red, and white pills in the morning before breakfast and at dinner." Or, "Doc., I'm not sure if it is a green pill or an orange pill, but I know I take one of those at bedtime."

This dialogue is the edge of a treacherous cavern that leads to a phenomenon called "polypharmacy." At worst, polypharmacy means the patient, doctor, and pharmacist have no idea what medications the patient takes. At best, the patient gets all his or her medications through a prescription benefits management supplier or single pharmacy, and then the pharmacist holds all the knowledge about drugs. A phone call to the pharmacy during the visit can occupy **10-15** minutes of valuable time, and is generally impractical although sometimes necessary.

Returning to the diabetes visit, it becomes clear that medication reconciliation is a critical component of the healthcare visit that can take anywhere from **5-10** minutes, depending on the organizational skills of the patient and the quality of the previously constructed list. In many instances I ask patients to bring in all their medications on the next visit so I can sort through their myriad drugs and clean up the medication reconciliation list.

Transparency to the time elements of a physician visit is important today because benchmarks are in place for all healthcare providers. Thus, a rapid visit means more money for all. Our estimates show that these three elements of the patient visit (introduction, prior medical record review, and medication reconciliation) account for **9 to 17** minutes. So much for the 15 to 20- minute follow-up patient visit. For a complex patient who might require 45 minutes, these foundation stones of care have occupied almost one-third of the follow-up visit time.

Getting Reimbursed

Then it comes to reimbursing the physician for the appointment. For the majority of first-visit patients, I use a time code since I don't fulfill the standard level four or level five coding rules using conventional auditing. I have established a simple macro shown below that allows me to validate the time spent with the patient:

Today I built a LIFEMAP for my patient and instructed her on proper self-blood glucose monitoring (SBGM). The LIFEMAP is an innovative diabetes care tool that integrates an average day in the life of my patient with particular attention to meal and snack times. This tool is explained to the patient and then used as a template upon which self-blood glucose monitoring is superimposed. The patient receives a logbook and is instructed to chart SBGM levels at the indicated times that I have written into the book.

The LIFEMAP allows me to obtain useful information on the follow-up visit to best-prescribed medications that will lead to better blood sugar control. Today I spent ___ minutes building the LIFEMAP and explaining the timing of SBGM. I spent more than 1/2 of ___ minutes giving my patient the above-stated counseling and coordination of diabetes care attested for in this note.

This macro gives me the freedom to implement quality diabetes care that is focused on high-impact interventions.

I find little pushback once the insurance company reviews these time codes. They are very specific to the interaction with the patient and help the processing insurance company understand exactly how we spent our time together in the office. This serves my time, the patients, and my overall healthcare delivery. It is win/win/win. My time is valued and reconciled, the patient gets a better level of care, and I don't have to fight with the insurance company to demonstrate how I spent my time with the patient. I predict we will continue to move to a payment model for value-based care in the near future. As we do, our cloud-based LIFEMAP platform will gain further value by allowing healthcare providers to practice at the top of the license and deliver outstanding diabetes care using digital patient mobile tools that inform a cloud-based data warehouse and prescribing algorithm. In essence, this will disrupt the present model such that the patient will have fluid access to the healthcare provider and system. The data will be visible to both patient and provider, and the provider will be able to deliver outstanding care with algorithm-based recommendations and bill for time spent in this new endeavor. Most importantly, the patient will have better blood sugar control and better health. The diagram below summarizes the disruptive transition in chronic care delivery from its present 100-year old status to a modern 21st century care delivery model:

Present Rubric for Healthcare Providers	LIFEMAP for Healthcare Providers/Patients
Provider-centered model	Patient-centered personalized data-driven model
Office visit payment model (fee for service)	Office visit/telemedicine payment model (pay for performance)
High- and low-impact activities mixed together	High-impact activities only
Old model of history, physical exam	Build a LIFEMAP
Assessment	Assign SBGM* rubric
Healthcare provider assigned treatment plan	Data-driven treatment plan

*Self-blood glucose monitoring

The Chronic Care Model, or any model for that matter, is not a recipe for success. Since the unique toll diabetes takes on your body is often specific to each patient, we as healthcare providers have to use our time in the office wisely to offer up the same level of flexible and unique care. Should we do so, we can begin to better control the patient and his or her conditions without wasting their time or ours. In the next three chapters, I will outline exactly how these initial visits in this model should look.

Chapter 9

A Trio of Treatment

The LIFEMAP offers both doctor and patient a tremendous opportunity to provide and receive top-notch diabetes care that is thoughtful, tailored, and uniquely differentiated for each patient. It is almost unheard of in our time-constrained world of medical diagnosis and treatment to think that a doctor is offering patients carefully plotted-out treatment plans that account for their own unique symptoms and challenges, rather than treating the general causes and effects of a specific disease. A more specific and focused approach at the ground level can help providers deliver better overall care to their patients. The LIFEMAP assumes at least three office-based visits with the patient, each building on the last and implementing a greater level of care.

Each of these visits provides patients what I have termed, "foundation stones of diabetes care." It is remarkable to discover that most individuals with diabetes don't understand simple principles of glucose production from the liver, post meal hyperglycemia, why to check blood sugar at home, and treatment strategies. It is also exciting to witness the transformation from helpless patient to empowered partner when the patient finally gets it and understands the principles of the LIFEMAP. This transformation changes the diabetes equation from you (the health-care provider) do it to me, to we do it together. Once the patient has connected with the LIFEMAP all good things happen and now you, the patient, control your diabetes instead of the other way around. This chapter will not just outline each of the visits, but also discuss what the healthcare provider will work to accomplish with each bite-size interaction for a patient with diabetes.

The First Visit

The first visit is special because it creates an opportunity to build trust and collaboration between patient and physician. Most patients with diabetes come to the office thinking to themselves that the doctor knows best. However, this is only half true. What we've learned over time is that diabetes is a data-driven disease, because the goal of treatment is to normalize blood glucose levels. Thus, we must immediately break the common misconception among patients with diabetes that the doctor knows best. The truth of the matter is that the healthcare provider is only as good at treating as the analytics she has in front of her.

Therefore, it is important to understand that the doctor doesn't have magical numbers for insulin dosing or medication prescribing that float down from heaven. Since greater than 99% of diabetes care takes place at home or at work, it is important for the doctor to empower the patient and explain that diabetes management is a shared responsibility. A doctor can only do so much. She can create a thoughtful treatment plan, but in the end the patient has to execute it if the two want to secure shared success. The conventional acute care model of chief complaint, history of present illness, past medical history, etc., and physical exam must be transformed into a chronic care model that changes the conventional paradigm we are taught in medical school to one of **high-impact** interventions only. We will talk more about this later, but for now the first visit should focus on social determinants of disease, medications, and building the **LIFEMAP**. In addition, the doctor should address any acute-focused complaints addressed by the patient. If not, a general physical exam is a waste of time.

On the first visit it is important to spend time with the patient explaining the four critical time points for glucose control. Remember, these include before breakfast and two hours after each of the three meals. Sometimes individuals only eat two meals a day so the LIFEMAP has to be configured for that lifestyle. Doctors should also explain that the liver is a sugar factory that causes high blood sugar in the morning, even when you don't eat food during the night. These talking points reinforce the reason why you as the doctor can then request that the patient checks his or her blood sugar four times daily. One exception to this rule is for patients with insulin-dependent diabetes and labile blood sugar (blood sugar that fluctuates widely). In those cases, patients should complete at least seven blood sugar checks per day. Otherwise, for most patients with typical type 2 diabetes, it is a relief to your patient when she hears that blood sugar checks need only be done two or three days per week. Since Medicare and Medicaid provide one glucose test strip daily for patients not on insulin and three strips daily for those on insulin, it is important to remind the patient on oral agents that they might not have enough strips to complete seven tests per day. Let them know it is okay not to test every day because useful information will be obtained to assess the trend of blood sugar over weeks and months and ultimately keep their blood sugar under excellent control. Ask the patient that they maintain a four-times-daily regimen until the strips run out.

Assuming they accomplish this task from the first visit to the next, on the second visit you, as their healthcare provider, will have a nice dataset to interrogate, despite the limitations Medicare and Medicaid create. The centerpiece of the first visit is actually building the **LIFEMAP** for your patients, so make sure you spend the majority of your time doing so. Spending time discussing nutrition on the first visit complicates the instructions and confuses the patient. Your best bet is to leave nutritional guidance for a later date. However, those institutions or practices with a real-time nutritionist might consider having a dietary visit on the same day to create a nutritional inventory and explain the concept of carbohydrate consumption. It really depends on the practice and the resources or limitations in place.

If necessary, then I typically ask my patients if they drink any sugar-containing beverages, iced tea, sports drinks, juice, or milk. Since these are high carbohydrate foods rapidly absorbed in the digestive tract, I ask if they can decrease or eliminate them from daily intake. This is often a serious task for individuals because many sugar-containing beverages (especially those with high fructose corn syrup) are highly addictive. If I give nutritional advice on the first visit, then this is all that I discuss. It is a big lifestyle transformation for most patients.

Next, I ask about social issues.

The goal at the end of the first visit is to have accomplished the following high-impact activities:

- Determine the duration of diabetes and microvascular and/or macro vascular complications.
- Determine social determinants of disease.
- Medication reconciliation.
- Build **LIFEMAP.**
- Explain the dynamics of insulin secretion from a healthy pancreas.
- Explain the logic for SBGM three days per week, four times daily.
- Eliminate or decrease intake of sugar-containing beverages from diet.

- Address acute complaints with limited physical exam.
- Follow-up visit in four to five weeks.
- Assuming you do so, you will prepare your patient for progressing into the next visit.

The Second Visit

Patients should schedule their second visit to occur four to five weeks after the initial one. The second visit is often where the doctor should initiate the first treatment step for the patient. Unless you need to manage seriously uncontrolled blood sugar (typically with values greater than 400 mg/dl), there is time to obtain a useful dataset and hold off treatment. When you have a case where you are concerned by a patient's high blood sugar (or when hemoglobin A1C is >10%), recommend a dose of long-acting bedtime insulin in accordance with standard guidelines.[40] If your patient has created a logbook with a trend analysis or active management profile, you should now have home glucose data that will help to inform treatment decisions. Assuming the primary care provider has already initiated anti-diabetic treatment (e.g., metformin, sulfonylurea, with or without DPP-IV inhibitor), the healthcare provider can then make an informed treatment decision based on the glucose logbook. Soon, we will have a cloud-based platform that will allow patients to record self-blood glucose monitoring directly over their smartphone into a high-security, HIPAA-compliant, cloud-based diabetes platform. The healthcare provider can then pull up the blood glucose results directly on his/her computer system or smartphone and build a treatment strategy right there. One simple case scenario is in the event of elevated fasting blood glucose. There are several treatment options for this case including:

- Instituting or increasing bedtime insulin: This is targeted therapy to control fasting blood sugar since long-acting bedtime

---

40 American Diabetes Association, "Standards of Medical Care in Diabetes," *Diabetes Care,* 2018, 41, Suppl 1, s1-172.

insulin lasts up to 20-30 hours and keeps the liver under control during sleep.

- Adding SGLT-2 inhibitor: This is a general approach to off-load sugar into the urine, thereby lowering blood sugar throughout the day and night. Excessive urination is one drawback to this treatment.

- Increasing the dose of existing antidiabetic medications: This is a shotgun approach to treatment with a goal of lowering blood sugar simply by raising medication dosing.

- Alternatively, the elevation of post meal glucose suggests a form of insulin-deficient diabetes with or without insulin resistance. In this case, the pancreas is unable to control post meal glucose and/or hepatic glucose production.[41] Here again, there are several treatment options that include:

- Initiate saccharidase inhibitor with meals: This treatment attempts to block intestinal absorption of sugars and carbohydrates after meals.

- Begin pre-meal insulin: This is targeted therapy to control post meal blood sugar directly by adding insulin.

- Add incretin mimetic or SGLT-2 inhibitor: These are shotgun treatments that can be tried to see if they yield better glucose control after meals. They do not directly address the problem of insufficient insulin production, but work through indirect mechanisms.

In all cases, the logbook profile should drive treatment decisions and follow the rule of trying to recapitulate normal insulin action to control the glucose profile. This approach does not assume any particular treatment, but lets the data drive the decision making. If the goal is to

---

41 R. A. Rizza, "Pathogenesis of Fasting and Postprandial Hyperglycemia in Type 2 Diabetes: Implications for Therapy," *Diabetes*, 2010, 59(11), 2697-2707.

normalize blood sugar, then the healthcare provider should use whatever strategy is necessary to achieve this goal without bias. In the case of SGLT-2 inhibitors, the normal physiology is somewhat violated because glucose is off-loaded into the urine.

SGLT-2 inhibitors block a protein in the kidney that causes sugar to be drawn back into the bloodstream. Since sugar tends to enter the urine when elevated in diabetes, it would be good to eliminate this sugar in the urine. Unfortunately, the kidney is smart and reabsorbs about 90% of this sugar unless a drug like SGLT-2 inhibitor is there to block this reabsorption.

In this case, the healthcare provider should take careful consideration when using SGLT-2 inhibitors with insulin. The reason for doing so is that severe hypoglycemia can occur in individuals with impaired kidney function. Also, additional caution needs to be taken when using this class of medications with type 1 diabetes (see package insert for full explanation).

An additional component of the second visit is filling in the gaps that occurred on the first visit.

Here, the treating physician can obtain information about sleep habits, obesity, dietary choices in a general manner, family history, and aspects of the physical exam that were not previously addressed. Several key questions are as follows:

- Do you have insomnia? How many hours per night do you sleep?
- Did you gain weight recently or have you struggled with this before?
- Is there family history of diabetes in your parents or grandparents?
- Do you tend to snack after dinner or during the night?
- Do you have depression?
- Are there issues we need to discuss that have not been addressed previously (e.g., lumbosacral disc disease, varicose veins, arthritis, etc.)?

Often, these secondary factors become the dominant behavioral drivers of disease management.

For example, an obese patient with lumbosacral disc disease will not have the ability to exercise. This patient might have a chronic pain syndrome, depression, and prioritize glucose control as a secondary health risk. A patient with chronic insomnia might need a sleep study to identify sleep apnea or might have insulin resistance due to sleep deprivation with loss of normal circadian rhythms. These factors, if unattended, will prevent optimal glycemic control and impede efforts to manage diabetes.

The second visit offers the healthcare provider an opportunity to assess patient compliance. Sadly, patients often forget their logbooks, misunderstand the instructions given on the first visit, or lack financial and/or social resources to achieve compliance. It is important to recognize these social determinants because if unaddressed, they will lead to a scenario of future noncompliance and treatment failure. It is also important to recognize that small steps can lead to a virtuous outcome for diabetes care if the patient and the healthcare provider can agree on a set of goals. This might translate to improving A1C from 11.0% to 9.5%, a big win for both patient and physician.

As a wrap-up to the second visit, the healthcare provider should reinforce self-blood glucose monitoring (SBGM) four times daily with a weekly frequency (twice or three times weekly) agreed upon by the patient and an area of nutritional focus for the patient. In the near future, we will have smartphone tools to address many of these gaps created by simply forgetting to check blood sugar at appropriate times and forgetting to enter numbers. Perhaps the patient will try to eliminate juice and soda from his diet; perhaps the patient can be more attentive to carbohydrate consumption and eat smaller portions; perhaps the patient recognizes the logbook is a critical disease management tool that gives him control of his diabetes. Any of these action items becomes a focus for the third visit and provides foundation stones for virtuous diabetes care.

The goal at the end of the second visit is to have accomplished the following high-impact activities:

- Identify hot spots where blood sugar is highest.
- Implement a personalized treatment plan to treat the hot spots.
- Identify barriers to care that might impede progress.
- Obtain lab tests to benchmark improvement in blood sugar control.
- Identify other diabetic risks related to microvascular and macro vascular disease.
- Agree upon goals of care and adapt LIFEMAP if necessary.

The Third Visit

The third visit is the most critical visit of the set. It is the first time both patient and physician can see the pieces of diabetes care coming together. The LIFEMAP is again the centerpiece of the visit because it allows the healthcare provider and patient to visually see the improvement in blood sugar control at each of the four critical time points. The LIFEMAP can now be easily and rapidly interrogated and compared from the second to third visit. Blood sugar trend analysis can provide a rapid view of the next step for treatment, as well as offer the physician and patient confidence that glucose control is actually improving. The third visit gives the provider an opportunity to fill in previously omitted gaps. For example, there is ample time in the third visit to explore family history, detailed nutritional history, comorbidities, and social determinants of disease because the hard work of building a personalized diabetes treatment plan, a LIFMAP, has already been achieved.

The third visit is also important for the patient, who should start to recognize the underlying strategy to hit their blood sugar target. Patients often comment about how they like having days off from blood sugar monitoring when using trend analysis. For those who are on an active management protocol, they recognize that much of the variance in blood sugar control is as or more important than absolute numbers. As we learn from numerous studies,[42,43,44] hypoglycemia is a serious concern not just for individuals with type 1 diabetes, but also for those with type 2 diabetes. Hypoglycemia can lead to mild symptoms of feeling woozy, headache, dizziness, sweating, heart palpitations, and cognitive impairment. Severe hypoglycemia can lead to hospitalization, and even coma, so there is real concern to maintain blood sugar in a controlled, but safe range (~100-150 mg/dl).

A number of factors can increase the risk of low blood sugar, including increased blood sugar variance and overly aggressive therapy. When blood sugar is more tightly controlled (trying to get the blood sugar ever lower) and the numbers get lower, then the risk of hypoglycemia increases. New technologies like continuous glucose monitoring devices have improved detection and prevention of hypoglycemia, but the risk of hypoglycemia is real when you have diabetes, and even with the best efforts, is unavoidable in most patients.

Therefore, a key factor in maintaining good glucose control and therapeutic compliance is mitigating hypoglycemia. The majority of patients on active management protocols are using multiple daily insulin injections or an insulin pump. These individuals require detailed analysis of

---

42 W. H. Polonsky, L. Fisher, D. Hessler, and S. V. Edelman, "Identifying the worries and concerns about hypoglycemia in adults with type 2 diabetes," *Journal of Diabetes and its Complications*, 2015, 29(8):1171-1176.

43 J. Grammes, M. Schäfer, A. Benecke, et al., "Fear of hypoglycemia in patients with type 2 diabetes: The role of interoceptive accuracy and prior episodes of hypoglycemia," *Journal of Psychosomatic Research*, 2018, 105:58-63.

44 E. B. Schroeder, S. Xu, G. K. Goodrich, et al., "Predicting the 6-month risk of severe hypoglycemia among adults with diabetes: Development and external validation of a prediction model," *Journal of Diabetes and its Complications*, 2017, 31(7):1158-1163.

the glucose log to pick up trends in hyper- or hypoglycemia. Often, small tweaks can significantly impact the overall glucose control. Since 7.0% is my A1C target (~150 mg/dl average blood sugar over 3 months) for patients on active management, there is generally room to increase the intensity of management without significantly increasing the risk of a hypoglycemic episode.

In addition, for those individuals on a four-shot insulin regimen with basal insulin at bedtime, it is prudent to build a basal insulin scale to decrease the risk of hypoglycemia overnight. Since there is typically no discussion of dosing basal insulin with a scale, most patients will adjust their bedtime insulin dose without telling their physician. It is peculiar that the prescribing healthcare provider would think that 20 units of a long-acting insulin at 10 p.m. is adequate for a blood glucose level of both 105 mg/dl and 185 mg/dl. This is illogical and incorrect. The basal insulin output from the pancreatic beta cells would naturally self-regulate to those two ambient glucose concentrations. Therefore, I build a basal scale (long-acting insulin) proactively with my patient that might look something like this (according to ±10 p.m. glucose level): 100-150 15 units; 151-200 20 units; 201-250 25 units; 251-300 30 units; 301-350 35 units; 351-400 40 units. Note: Please do not use these actual doses for patients. The basal bedtime insulin dosing scale must be individualized for each particular patient. The basal dose schema presented here is just an example of how this might be constructed.

The doctor can then adjust the basal scale on future visits according to fasting glucose information and self-reporting from the patient as to how many hypoglycemic events occurred during the previous visit interval. Here, the goal is to adjust the bedtime basal scale to optimize the blood sugar in the morning before eating food and to minimize the episodes of hypoglycemia (low blood sugar) overnight. The goal at the end of the third visit is to have accomplished the following high-impact activities:

- Review improvement in average glucose and estimated A1C between visits.
- Rework LIFEMAP, if necessary, to fine-tune glucose monitoring times (datasets) with particular attention to uncontrolled hyperglycemia and/or episodes of hypoglycemia.
- Fill in gaps for physical exam (if necessary) and social determinants of disease.
- Adjust treatment according to LIFEMAP datasets to improve glucose control.

While there will obviously be more visits to come, and additional healthcare provider/patient interactions, these first three visits are crucial in implementing the LIFEMAP and a tailored structure to medical treatment for diabetes. Should the healthcare provider take the time to truly walk his or her patient through these initial visits with a goal of achieving the high-impact activities outlined above, then the patient will greatly benefit from this carefully tailored and ground-level healthcare treatment plan. For many, three visits will set them on the path to successful diabetes management.

In addition, it is my aspiration that our cloud-based LIFEMAP will bridge the gap between office visits by allowing patient and healthcare provider to seamlessly interact in real time through smartphone technology to optimize diabetes management and blood sugar control.

Chapter 10

Chronic Diabetes Management

I n the previous chapter, we outlined an innovative approach to diabetes care using the LIFEMAP as a centerpiece. It should be clear
to both healthcare provider and patient that this is not your typical doctor office visit. We have revamped an approach to medical care
that has its roots in a rubric more than 100 years old. How did we end
up with a 100-year-old approach to medical care in the 21st century? It
is a short story about the history of medicine and how things rapidly
changed over the past 30 years.

In the distant past during the great era of the anatomists, knowledge
about the workings of the body was limited to dissection of cadavers and
description of the various organs in the body. Understanding how these
organs worked took many decades to figure out. Patients with acute disease presented to hospitals for archaic treatments that did not work (e.g.,

bloodletting, trepanning, cupping). The poor souls subjected to these techniques were examined by a core of physicians and most often sent home to die. Since diagnosis was inaccurate and treatments limited, most individuals succumbed to acute manifestations of chronic illness (e.g., tuberculosis, cancer) or acute death (e.g., internal bleeding, heart attacks, or strokes). What was there to do for the patient other than collect a lot of information about their lifestyle, family history, life habits, and other such information?

As the modern era of medical physiology emerged in the mid-20ᵗʰ century, physicians now had tools to perform measurements which led to more accurate diagnoses like hypertension, diabetes, high cholesterol, asthma, and the like. Drugs to treat these conditions followed. Throughout these two great eras of medicine a rubric for diagnosis and treatment emerged called the History, Physical Exam, Assessment, and Plan. This approach is still important today in the acute hospital setting where a sick patient comes to receive urgent care that requires an accurate diagnosis and treatment plan. Here, the small details of past medical history, family history, travel history, medications, allergies, and a thorough physical exam all provide important clues to the healthcare team to make an accurate diagnosis and apply appropriate treatment. However, something changed in the last 30 years, the emergence of the chronic disease state.

Medicine evolved rapidly during this time period. Diagnosis became more precise and treatments more effective. Instead of patients being sent home to die, they now had chronic manageable disease they could live with for decades. A good example of this transition is human immuno-deficiency virus (HIV). Previously, HIV led to AIDS (acquired immune deficiency syndrome) and death. Now HIV is a treatable disease that requires chronic medication to suppress the virus. We now monitor viral load in the blood and side effects of the medications instead of watching an acute viral illness lead to death. The focus of quality for HIV patients has shifted from doing nothing and waiting for the worst to happen to monitoring the suppression of the HIV virus and assessing the patient for secondary effects from the medications.

So too, diabetes was once a disease that could be managed with only two drugs, insulin and oral sulfonylurea (insulin stimulating) agents using crude tools to measure blood glucose in the hospital. Chronic care for diabetes didn't exist because we didn't have the technology necessary to obtain home glucose monitoring. Insulin was typically administered in a hospital clinic because needles were not available for home use. Once the transition occurred from an acute care model for diabetes to a chronic care model, one would have expected that the rubric for disease management would have changed. But it didn't. We transposed the same acute care model of History, Physical, Assessment, and Plan to the chronic care setting without realizing we needed to pivot. The acute care model all of us as healthcare providers grew up with is broken in the chronic care setting. Activities that were routinely valued in the acute care setting as high impact became low-impact activities in the chronic care setting, but we didn't realize it.

Over the years we have learned that those patients suffering from chronic diseases, like diabetes, require a different set of skills and tools than those with acute illness that requires intensive hospital-based care. It's apparent the medical community failed to recognize this difference and therefore disease management for acute and chronic care continues to be lumped together in medical school training. This has unfortunately cascaded down to Medicare reimbursement and electronic medical record documentation where healthcare providers are forced to use the same templates and rubrics for both acute and chronic care delivery.

As you might imagine, this creates a dangerous environment for those patients with chronic symptoms of diabetes. They need intensive treatment and assistance, but focused on different endpoints of care than a patient with diabetes that needs acute hospital care. For example, a patient with decompensated diabetes and uncontrolled blood sugar might be admitted to a hospital ward for dehydration, acute kidney injury, and life-threatening acid buildup in the blood that can lead to irreversible organ damage. The healthcare provider approaches this patient very differently from an individual with long-term poorly controlled diabetes who suffers from chronic manifestations of nerve disease (diabetic neuropathy), kidney disease (diabetic nephropathy), or eye disease (diabetic retinopathy).

Therefore, it is time to formally recognize and codify a chronic care model for diabetes that is very different from the acute care model I support robustly for diabetic patients who require acute hospital-based diagnosis and care. These two models are not mutually exclusive once we recognize there are significant differences between acute and chronic healthcare delivery. The LIFEMAP is an attempt to codify a high-quality diabetes management platform designed for chronic outpatient diabetes care; the place where 99% of diabetes care takes place.

The good news is that the medical industry recognizes these important distinctions, and changes are afoot. When approached with great logic, while history, physical, assessment, and medical decision making are the cornerstones of acute disease management (we need to interrogate and solve the problem), chronic disease management is more about building foundation stones of care for continuity and prevention. When viewed in this light, certain critical elements of the acute care model fall away while new components of medical intervention and counseling become relevant. We know this intuitively as physicians and healthcare providers, yet we cling to an antiquated model of one-size-fits-all care that is entrenched in the 20th century. This leads to poor treatment for most, especially those with the greatest needs. The additional pressure of medical productivity and time management mandates that we do things differently in the future, starting now.

The Challenge in Chronic Care Management

Diabetes care is a prototype of chronic disease management. In 25 years of practice, I recall rare instances where physical examination aided me in my care plan. This is especially true when a patient lacks a specific chief complaint. What are we to examine? Obviously, focused chief complaints must be evaluated and assessed in the chronic care environment. Examples might include statements/responses from patients like: my feet are numb at night = foot exam for neuropathy; or I have shooting pain down my left leg = palpation of lumbosacral region for point tenderness and X-ray of lumbar spine; or chest pain on exertion = coronary (heart) evaluation.

When it comes to chronic disease management, the physician exams the patient in response to the patient telling him or her what hurts. The physician then interrogates the problem. If the patient has no complaint, then you are in standard chronic care mode, and a physical exam will generally yield nothing in terms of direction and treatment. Alternatively, a physician can make many diagnoses when he or she pays attention to the patient complaint in the chronic care setting. The healthcare provider who delivers a shotgun physical exam on every visit would otherwise ignore this writing on the wall. Recent evidence-based studies of the annual physical exam in asymptomatic individuals suggest a lack of utility.[45] Therefore, in most instances a physician's physical exam for chronic diabetes management becomes a low-impact intervention for patients.

With that in mind, what are the high-impact interventions for the diabetes chronic care model? Simply stated, to control the blood sugar.

---

45 H. E. Bloomfield, T. J. Wilt, "Evidence Brief: Role of the Annual Comprehensive Physical Examination in the Asymptomatic Adult," *Department of Veterans Affairs Health Services Research & Developmental Service*, 2011, retrieved from https://www.ncbi.nlm.nih.gov/books/NBK82767/pdf/Bookshelf_NBK82767.pdf.

When referencing the LIFEMAP, we built the components of this model from the bottom up to focus on blood glucose control.

We start by building the foundation for a strong standard of care. This includes:

✓ A discussion about the four critical time points during the day when we measure home glucose (before breakfast + 2 hours after each meal).

✓ The nature of trend analysis that doesn't require glucose monitoring every day (at least for noninsulin-Xgion for point tenderness and xrequiring diabetes).

✓ The timing of insulin action.

✓ A simple discussion about foods rich in carbohydrates.

✓ The plan of attack to control blood sugar.

It is remarkable how very few patients understand these foundation stones. Therefore, they are often misinformed about why they are performing certain diabetes-related activities. A good example of this is a conversation I often have with my new patients concerning their home glucose monitoring.

I start by asking my patient how often they check their blood sugar.

"Twice a day," they frequently respond to me.

I ask, "What time of day?"

The inevitable answer is, "Before breakfast and before dinner."

I then ask the patient if they bring their logbook to the doctor.

They most often say, "Yes."

Next, I ask what the doctor does with this crucial information.

Of course, the answer is, "He looks at the logbook."

"And then what happens?" I ask.

The patient is stymied now and says, "He tells me to keep checking my blood sugar."

This dialogue occurs about 80-90% of the time I see new patients. The patient has no idea why they check their blood sugar or what the

doctor actually does with this information. There is a common disconnect between an insufficient dataset and the inability to build a treatment
plan around the dataset. That then leads the doctor to offer a bottled
version of healthcare to those patients with diabetes.

Diabetes is a data-driven disease that relies on a quality set of glucose
values in the home setting. As mentioned before, this method of interro-
gating home glucose is critical to overall success (A1C reduction). One
example of time well spent in the chronic diabetes treatment paradigm
is offering the patient intimate knowledge of the reason for performing home glucose at critical time points. The LIFEMAP reinforces these
principles and provides structure to all healthcare providers so they have
a uniform playing field to integrate their treatment plans.

That leads us to a very important question: How should it be done?
If we are to separate acute care from chronic care, we have to do so with
great thought and a flexible yet detailed approach. Asking general and
canned questions doesn't serve all patients at the same level. In fact, it is a
disservice to some. However, as you might imagine, we can in fact sculpt
a level of care that meets the needs of all types of diabetes patients. That
is where the LIFEMAP comes in.

Build a LIFEMAP so you and the patient connect on the critical
time points to check glucose at home. Tell the patient to check SBGM
3 days per week, four times daily, at those designated times (remember,
everybody has a different LIFEMAP, so the time for glucose monitoring
will often be different for different individuals). Find out if the patient
typically skips meals and if so, how often. We will see more of this when
we come to the section on Case Studies. Ask your patient to write down
their numbers in a logbook or on a sheet of paper. Have the patient
return to your office with their SBGM in 4-5 weeks so there is adequate
data to interrogate.

This type of individualized care is helpful to combat even the most
serious forms of diabetes. After doing so, we can look at the gaps in

care that arise from using trend analysis. Simply put, while looking for trends in glucose patterns information gaps arise because we are limiting the dataset. Continuous glucose monitoring devices can easily overcome these data gaps, but for most individuals who are pricking their finger a limited number of times per week there are points of information (glucose values) that could be useful but were not obtained. In the next chapter, we will learn how to recognize and navigate these data gaps and fill them with information that moves the diabetes care and treatment forward for the patient. Sometimes understanding the limitations of what you cannot do for a patient is as important as what you can do. The LIFE-MAP can be adjusted to the needs of the patient and data gaps can be filled when critical information is necessary, but not available at present.

Chapter 11

Recognizing Gaps in Care

A s we work towards improving chronic diabetes management in patients, it is extremely important to maintain a consistent and in-depth understanding of the gaps that occur in patient care. The LIFEMAP provides healthcare provider and patient with a neat rubric for improving blood sugar control. It opens up the conversation so we can replace a one-size-fits-all approach to diabetes care with a much more insightful and specifically tailored treatment program. In the end, personalizing diabetes care might be the difference between a healthy patient and an unhealthy one. But even the LIFEMAP isn't perfect, and this chapter will outline the gaps in its care and in the dataset.

As we secure a better understanding of these gaps, we can then achieve high-quality diabetes care. Most doctors offer the following response to patients in need: "Check your blood sugar more frequently." In many ways this has become the default answer of sorts. However, there are numerous limitations on glucose test strips by CMS, and a frequent distain for finger pricking. Thus, we must develop a set of "work-arounds" to achieve success.

First, we need to clearly define a "gap." Gaps occur when the healthcare provider doesn't have enough information to fully understand the ups and downs of blood sugar in a particular patient. This will become much more obvious as we go through different examples of data gaps, but here is one simple example.

A patient returns to your office with a well-constructed LIFEMAP. Blood sugars are recorded before breakfast and two hours after each meal. You, the healthcare provider, recognize that on certain mornings before breakfast the blood sugar is 100-140 mg/dl, excellent numbers. On other mornings, the blood sugar is 55-75 mg/dl, dangerously low, perhaps. Your patient is taking a fixed dose of 20 units of long-acting insulin at bedtime and you realize it is possible on certain nights before bedtime the blood sugar starts off at 108 mg/dl, for example, and on other nights before bedtime the blood sugar starts off at 168 mg/dl, but you don't have that information on hand.

This is an example of a data gap—missing information that would be critical in managing the blood sugar for the patient. Perhaps the fixed dose of 20 units of insulin at bedtime is the right dose when the blood sugar is 168 mg/dl. Perhaps the fixed dose of 20 units is too much when the blood sugar is 108 mg/dl.

Therefore, recognizing the data gap, you ask the patient if she is willing to check her blood sugar every night at bedtime and every morning before breakfast to obtain useful information to fill the data gap. On the next visit when the patient returns with new LIFEMAP data you see that your hypothesis was correct. When the blood sugar at bedtime is below 120 mg/dl, 20 units of long-acting insulin at bedtime might be too much and could cause low blood sugar in the morning or overnight. When the blood sugar is above 150 mg/dl at bedtime, 20 units of long-acting insulin is just right. Next you build a bedtime scale for long-acting insulin so the blood sugar is well controlled almost always in the morning. You have just filled a data gap in the LIFEMAP.

The First Gap: High Blood Sugar After Meals

We see another example of this data gap in the patient who has well-controlled fasting glucose, but also post-prandial hyperglycemia and uses insulin. Perhaps the patient is on a fixed-dose four-shot insulin regimen with long-acting insulin, 25 units at bedtime, and recombinant short-acting insulin, 15 units three times daily before meals. When the patient goes to the doctor, he or she brings in a neatly crafted glucose log three days per week, four times daily, that shows typical blood sugar after dinner between 150-320 mg/dl.

As a healthcare provider, you might discuss decreasing the carbohydrate content of his or her dinner, but you lack a glucose measurement before dinner to fully comprehend the resultant post meal hyperglycemia. We have to change at least one element of the LIFEMAP in this circumstance, and I would recommend removing blood sugar monitoring before breakfast.

We can then reconfigure the rubric to check blood sugar after breakfast, after lunch, before dinner, and after dinner. This dataset gives the healthcare provider an opportunity to build a bolus scale for insulin dosing before dinner. While not a perfect solution, this approach gives patient and healthcare providers alike the opportunity to interrogate blood sugar dynamics after dinner and hopefully build an insulin-dosing scale that leads to better control. It is also important to discuss what will be done on the off days when pre-dinner glucose is not checked. For this situation I typically recommend picking an intermediate dose based on bolus scale usage and stick with that for the near future.

Having access to home glucose values at the correct time can make a big difference in the well-being of an individual and diabetes specifically. Another option for the patient above is to add a medication called saccharidase inhibitor before dinner. While not an answer for all individuals with diabetes and elevated post dinner glucose, this class of drugs blocks the absorption of carbohydrates in the intestine and might lead to better blood sugar control after dinner.

Sometimes, it is worth trying a drug like Acarbose or Miglitol (saccharidase inhibitors) before dinner to see if post dinner blood sugar can be better regulated. These decisions about fine-tuning blood sugar can only be accomplished with a LIFEMAP that gives the patient and provider a way to test and evaluate a particular intervention in a personalized fashion.

The Second Gap: The Problem with Long-Acting Bedtime Insulin

A second gap is one I have previously discussed. We recognize the patient who complains of nighttime hypoglycemia. Here for example, the patient is on a three-shot insulin regimen with mixed insulin, (e.g., Humalog 75/25 or Novolog 70.30 insulin) 30 units before breakfast; plus, recombinant short-acting insulin, 15 units before dinner; and a fixed-dose long-acting insulin analog, 20 units at bedtime; or a patient on oral medications during the day and bedtime long-acting insulin. The patient notes waking up at 2 to 4 a.m. two to three times per week with symptoms of sweating, hunger, and blurred vision. She typically doesn't check her blood sugar when this happens, but on occasion finds a reading between 40-60 mg/dl. She drinks juice, the symptoms resolve, and she goes back to bed with ensuing hyperglycemia the next morning. This demonstrates the second gap, which is the bedtime fixed-dose insulin dilemma.

Pharmaceutical companies market their long-acting insulins as fixed-dose solutions for managing basal insulin replacement therapy. But our patients are smarter than that. It is logical and somewhat obvious that 20 units of basal insulin at bedtime is good when the bedtime glucose level is 150 mg/dl, but not when the bedtime glucose level is 86 mg/dl. Fixed-dose basal insulin is an approximation that frequently does not work!

Again, we need to adjust the LIFEMAP glucose monitoring to interrogate basal insulin dosing before bedtime. As previously discussed in chapter 3 under "The Myth of Bedtime Insulin," it is constructive to shift the blood glucose testing from two hours after dinner to bedtime. The next step is to build a basal insulin scale as previously shown. So what happens the other four days when bedtime glucose is not measured? Find a compromise midrange basal insulin dose when your patient is blind to her bedtime glucose level or give your patient the option of checking blood sugar every night before bedtime. The net result of this maneuver is better compliance with bedtime insulin dosing and less hypoglycemia. While this is not a perfect fix, the point of this maneuver is to round the edges off hyperglycemia (which occurs by not taking any insulin at bedtime) and make glucose control a safer process.

The Third Gap: The Shift Worker Dilemma

Another gap in the LIFEMAP occurs when the patient is a shift worker. The major challenge for these individuals is to figure out when to recommend finger-stick glucose checking. Your patient is a 58-year-old male who works a graveyard shift four days per week from 8 p.m. to 6 a.m. He has a small meal at 2:30 a.m. He then comes home from work, eats breakfast at 7:00 a.m., and goes to bed at 7:30 a.m. The patient wakes up at ~12:00 noon and does his house chores. He eats dinner before his shift at 5:00 p.m. Here, I recommend obtaining blood glucose at 4:30 a.m., 7:00 a.m., 5:00 p.m., and 7:00 p.m. This is somewhat arbitrary and should be adjusted on a case-by-case basis to obtain useful information that can be translated into action items.

Obtaining finger-stick glucose readings before and after dinner gives you the opportunity to lock in the post dinner glucose level. The other time points are somewhat arbitrary, but will allow you to interrogate blood sugar in the middle of the night after food and before bed. Configuring a treatment strategy for this particular situation is tricky because one needs to consider insulin resistance that occurs from sleep deprivation and disruption of circadian rhythms.[46]

Generally, I prescribe a basal insulin dose before work at 7:00 p.m., pre-meal insulin at 5:00 p.m., and then decide whether pre-meal insulin at 2:30 a.m. is even necessary. For oral agents, it is important to recognize that SGLT-2 inhibitors will necessitate frequent trips to the bathroom, which might not be compatible with the particular work environment or sleep in the morning, while long-acting sulfonylurea agents can lead to hypoglycemia if the job requires physical labor and missed meals.

The Forth Gap: False A1C Readings

Another example of LIFEMAP curiosities occurs when the healthcare provider cannot use A1C as a measuring tool. This situation occurs in patients with end-stage liver disease, hemoglobinopathies (genetic abnormalities with hemoglobin), anemia (although iron deficiency anemia in rare instances can raise A1C level),[47] chronic kidney disease, and certain drugs like anti-retroviral or immunosuppressive agents. Healthcare providers often miss this phenomenon when they tell the patient for many years their diabetes is well controlled. Here, the LIFEMAP becomes a lifesaver because it is the only tool available to interrogate blood sugar control. Patients have been given instructions to check their blood sugar before breakfast and before dinner so these reading reinforce confidence

---

46 J. Bass and J. S. Takahashi, "Circadian integration of metabolism and energetics," *Science*, 2010, 330(6009):1349-1354.

47 E. Coban, M. Ozdogan, and A. Timuragaoglu, "Effect of iron deficiency anemia on the levels of hemoglobin A1c in nondiabetic patients," *Acta Haematologica*, 2004, 112(3):126-128.

that blood sugar is controlled since these readings often reflect the low numbers during the day. Often, I tell these patients not to be surprised if they see post meal blood sugar levels in the 200-400 mg/dl range. The second visit for these individuals allows blood sugar control to be reassessed and treatment strategy reconfigured to achieve improved blood sugar control.

It is also the necessary confirmation that A1C cannot be used to assess 3-month glucose control in such individuals. With a cloud-based LIFEMAP we will deliver to healthcare providers an average glucose from LIFEMAP data and an estimated A1C that can be substituted for a standard A1C measured in the laboratory. This information can be a useful tool to quantify improved blood sugar control over time when other tools are not available.

One Final Gap: Diabetes Medications Can Fail Over Time

It is important to understand that gaps in care will crop up unexpectedly when you use the LIFEMAP as a guiding principle for diabetes management. As we have discussed, the LIFEMAP is a simplified approximation of the average day-to-day blood sugar variability that individuals with diabetes or even prediabetes experience. The use of trend analysis in the LIFEMAP means we are not collecting all the possible information required to map out a complete and exhaustive analysis of blood sugar control. Devices such as continuous glucose monitors give patient and healthcare provider a complete view of day-to-day blood sugar variance. Even with these newer devices it is still necessary to use the information in a thoughtful manner and transform the data into a treatment plan. Since not all individuals with diabetes need such intensive glucose monitoring, the LIFEMAP approach to diabetes care is the next best thing. While much discussion has focused on individuals who use insulin, similar gaps occur with patients on oral agents.

The LIFEMAP gives healthcare providers and patients alike an opportunity to interrogate blood sugar using a trend analysis approach as just discussed. Sometimes I ask my patients to check SBGM 1 to 2 times per week, 4 times daily, and keep a log to ensure they are stable and "on the railroad tracks."

It is easy to be lured into a sense of complacency when blood sugar control is good and stable. Too often, I find patients who had excellent blood sugar control, but then gradually slipped down a rabbit hole into uncontrolled blood sugar because they didn't have a preventive approach to monitoring their blood sugar week by week. Since diabetes is not a stable disease, it can get better or worse depending on many health, social and behavioral variables that are not so easily quantified. Therefore, the only real remedy to prevent this transition from well-controlled blood sugar to uncontrolled diabetes is a systematic approach to self-blood glucose monitoring, the LIFEMAP. This approach prevents patients from falling down the rabbit hole and spiraling into inadvertent uncontrolled blood sugar.

Gaps are a normal part of any treatment protocol. There is just no magic or curative bullet at play here. The LIFEMAP is no exception to this rule. However, it does account for many of the substantial issues we see in the general treatment of diabetes. A patient with diabetes has unique and case-specific challenges. While the LIFEMAP may not solve every single one of these issues, it goes a long way to put the patient in a place where she can live a relatively healthy and active lifestyle. Let's see how this unfolds in the next chapter, which offers numerous case studies for review.

Chapter 12

Case Studies

Perhaps one of the most effective ways to demonstrate the LIFE-MAP in motion is through a series of case examples. You will find nine different case studies, along with an outline of each visit to the patient's treating healthcare provider. The cases presented here are not real patients, but an amalgamation of the more than 1,000 individuals with diabetes I have cared for over the last several years. I used the LIFEMAP approach in each of these examples, and the details should show the LIFEMAP in motion and applied to complicated and unique cases. It is purely coincidental if any case has resemblance to a real patient. They simply reflect the common scenarios I have treated over the years. No medical records were used to craft these case studies. Let's look at each one at greater length.

Case 1
Uncontrolled Type 2 Diabetes with Neuropathy That Is Easy to Fix

Case 1: Visit 1

Juan is a 46-year-old Hispanic male with a history of diabetes spanning over seven years. He feels fine, but notes tingling in his feet at night. He is presently taking metformin (1000 mg) twice daily and glipizide (5 mg) twice daily. He is not on an ACE inhibitor or ARB and checks his blood sugar once daily in the morning with self-reported readings of 110-160 mg/dl.

He is not particularly careful with his diet, but he has eliminated simple sugars. However, he continues to eat high levels of carbohydrates at mealtimes. Juan has a strong family history of diabetes in his mother and aunts. He is concerned about his 8-year-old son and 12-year-old daughter. His wife is healthy and has no history of diabetes in her family. He does not smoke cigarettes or use drugs. He drinks one six-pack of beer over the weekend. Juan is taking metoprolol (25 mg) twice daily for hypertension and has no allergies to medications. His recent A1C was 10.3%. He denies hypoglycemic episodes, but elaborates that he has burning and aching feet daily.

On first evaluation, his vital signs are: BP 146/90, Pulse 88, Body Mass Index (BMI) 34. His foot exam (moderate impact because complaint is present, but likelihood of physical finding is low) is unremarkable.

My **LIFEMAP** for Juan looks like this:

Wake Up (WU): 6:00 a.m.

Breakfast (B): 7:00 a.m.

Lunch (L): 1:00 p.m.

Snack (SN): 3:00 p.m.

Dinner (D): 7:00 p.m.

Bedtime (BT): 10:30 p.m.

We then construct the following rubric for blood sugar monitoring based on his **LIFEMAP** (the snack will integrate into the post dinner blood sugar so I typically do not request finger-stick glucose after snacks):

SBGM:

Before Breakfast (BB): 6:00 a.m.

After Breakfast (AB): 9:00 a.m.

Before Lunch (BL):

After Lunch (AL): 3:00 p.m.

Before Dinner (BD):

After Dinner (AD) 9:00 p.m.

Bedtime (BT):

I explain to Juan he should not worry about his blood sugar results because we are just getting started. Rather, we need to obtain a useful dataset. I ask Juan to write down the numbers in a logbook that I provide or on a piece of paper with rows and columns. I also explain that these times are not precise, and he can enter a blood sugar reading under the 3:00 p.m. column even if he eats lunch at 1:30 p.m. and checks his post meal glucose at 3:15 p.m.

I explain that we are looking for ballpark numbers so we can build a treatment plan on the second visit. I also explain that if he forgets his logbook on the second visit we will need to go by memory, but that I will give him exactly the same instructions on the second visit until he remembers his logbook. I tell him not to fake the numbers, even if they are bad. I tell him he will start taking a new medication called ACE to protect his kidneys. I also ask him to consider cutting back on his beer consumption over the weekend because this might be contributing to his neuropathy in his feet.

Visit 1 Summary:

1. Obtain diabetes relevant history.

2. Medication reconciliation.

3. Build **LIFEMAP** and SBGM at home.

4. Optional foot exam.

5. Begin lisinopril 5 mg daily.

6. Cut back on beer consumption if possible.

7. Return to Care (RTC) in 4 to 5 weeks with logbook.

8. Visit time is 45 minutes; Level 4 new patient visit coding (99204).

Case 1: Visit 2

Juan returns in four weeks with a well-documented logbook. Here are the first three rows and last three rows of it:

LIFEMAP WITH SBGM*

Allergies none
Cigarettes none
ETOH 6 pk on wkend

BB 6:00 a.m.	AB 9:00 a.m.	BL.	AL 3:00 p.m.	BD	AD 9:00 p.m.	BT
163	245		310		225	
142	283		225		342	
188	227		288		310	
138	198		256		265	
127	188		224		256	
153	210		262		308	

Wake up　Breakfast　Snack ±　Lunch　Snack ±　Dinner　Snack ±　Bedtime

Time 6:00 7:00 8:00 9:00 10:00 11:00 12:00 noon 1:00 2:00 3:00 4:00 5:00 6:00 7:00 8:00 9:00 10:00 11:00 12:00 mn

Social determinants of disease   Avg. Glucose 230.5 Est. A1C 9.7%

☐ psychosocial ☐ financial ☐ access ☐ substance abuse ☐ compliance
Micro/Macrovascular Complications
☒ Neuropathy ☐ Retinopathy ☐ Microalbumin ☐ Macroalbumin
☐ CVD ☐ CVA ☐ PVD ☐ Carotid Disease
* Self blood glucose monitoring

He is feeling well, but complains of mild neuropathy at night. Today his BP is 138/84 and BMI 33.

He states that after seeing high blood sugar after meals he tried to cut back on his carbohydrates at mealtimes. He noticed that his numbers were significantly lower the last two weeks.

Juan states, "I would like to lose weight, but it's hard getting to the gym with my work schedule, and my wife cooks for me." I ask if it is possible to have his wife come in on one of his visits so she can sit with us and possibly meet with a nutritionist.

He agrees with my suggestion, but needs to check with her first. We look over his glucose log and I note that he has done a great job lowering his blood sugar on his own. I explain that while we could gain better control of fasting glucose with bedtime insulin, it would be more advantageous to try to improve diabetes control with an agent that is virtuous (one that does not cause weight gain in the context of lower glucose).

I explain to Juan that glipizide is a vicious drug in that it stimulates appetite and causes weight gain in lowering blood sugar.[48] We can use a newer agent called SGLT-2 inhibitor to control blood sugar and induce modest weight loss. I explain that this drug can cause blood pressure lowering, a good thing, as well as urinary tract infection, and genital yeast infection. We will also need to check kidney function before starting the medication. Since the profile for SGLT-2 inhibitors varies, I recommend empagliflozin (10 mg) daily in combination with his other medications with continued SBGM as before. Juan is excited to try this. I also recommend an eye examination since his last one was two years ago. He also repeats that his feet still hurt at night, but he has decreased his beer consumption to two beers on the weekend and this might lead to some improvement since alcohol can be toxic to the small nerves in his feet.

Visit 2 Summary:

1. Interrogate **LIFEMAP**.
2. Recommend nutritional consultation with wife.
3. Lab slip for basic metabolic profile and urine microalbumin to creatinine ratio to check kidney function.

---

48 S. E. Kahn, S. M. Haffner, M. A. Heise, et al., for the ADOPT Study Group, "Glycemic Durability of Rosiglitazone, Metformin, or Glyburide Monotherapy," *New England Journal of Medicine*, 2006, 355:2427-2443.

4. Referral to ophthalmology.
5. Begin empagliflozin 10 mg after kidney check and call or e-mail me.
6. Continue blood sugar monitoring via **LIFEMAP**.
7. RTC in 4 to 5 weeks.
8. Visit time is 30 minutes; Level 4 follow-up care visit coding (99214).

Case 1: Visit 3

Juan returns in four weeks with a logbook, normal renal function, a normal eye exam, and negative urine microalbumin to creatinine ratio. His wife has modified their diet and has reduced carbohydrates in their meals. He has been taking empagliflozin for the past two weeks with noticeable increased urination and lower blood sugar. His nighttime numbness in his feet has resolved. His blood pressure is 122/74 and BMI 32. We show his logbook values for the past two weeks below. I share with him that his estimated A1C is now 6.2% and his blood A1C will show significant reduction on the next measurement as he maintains excellent glucose control. He is continued on his medications without change and told to check SBGM two days per week; four times daily to maintain glycemic control. I increase lisinopril to (20 mg) daily, a kidney-protective dose. I ask him to follow up with me in three months and repeat A1C level.

Visit 3 Summary:
1. Interrogate **LIFEMAP**.
2. Maintain present medications and increase lisinopril.
3. Reduce SBGM to twice weekly.
4. Obtain A1C prior to next visit and check basic metabolic profile for elevated potassium from high dose lisinopril.
5. Visit time is 15 minutes; Level 3 follow-up care visit coding (99213).

LIFEMAP WITH SBGM*

Allergies none
Cigarettes none
ETOH 2 beers/wkend

BB 6:00 a.m.	AB 9:00 a.m.	BL.	AL 3:00 p.m.	BD	AD 9:00 p.m.	BT
105	135		155		143	
93	118		152		168	
116	133		109		154	
88	122		148		150	
122	156		136		150	
115	110		128		165	

Wake up ▸ Breakfast ▸ Snack ± ▸ Lunch ▸ Snack ± ▸ Dinner ▸ Snack ± ▸ Bedtime

Time 6:00 7:00 8:00 9:00 10:00 11:00 12:00 noon 1:00 2:00 3:00 4:00 5:00 6:00 7:00 8:00 9:00 10:00 11:00 12:00 mn

Social determinants of disease   Avg. Glucose 132 Est. A1C 6.2%

☐ psychosocial ☐ financial ☐ access ☐ substance abuse ☐ compliance
Micro/Macrovascular Complications

☐ Neuropathy   ☐ Retinopathy ☐ Microalbumin ☐ Macroalbumin

☐ CVD   ☐ CVA   ☐ PVD   ☐ Carotid Disease

* Self blood glucose monitoring

Case 1: Visit 4

Juan returns with new lab results, showing an A1C of 6.8%, which is a dramatic improvement from his starting point of 10.3%. His neuropathy is improved and he feels great. I recommend he continue with his LIFEMAP 1-2 times weekly to ensure he stays on track and follow up in three to four months for continuity of care. I tell him, "Your A1C reflects backwards three months so we might expect further reduction in the future if you keep on the same course." He is pleased.

Case 2
A Complex LIFEMAP, Insulin-Requiring Diabetes and Obesity

Case 2: Visit 1

Shauna is a 28-year-old African-American female recently hospitalized for diabetic ketoacidosis, a potentially deadly condition. She has had diabetes for about five years, ever since her second pregnancy. She was treated with insulin at that time and since then had been taking metformin (500 mg) twice daily and insulin.

She is not receiving diabetes care but follows up with her primary care provider intermittently. Upon discharge from the hospital, she was placed on lispro insulin 15 units three times daily with meals and glargine insulin 20 units at bedtime.

Her insurance does not cover glargine insulin so she has been taking only lispro twice daily with breakfast and dinner. She complains of blurred vision intermittently, vaginal itching with white discharge, and frequent urination three to four times during the night.

She has two children, a 7-year-old son and a 5-year-old daughter. Her grandmother helps as a care provider for her children. She is not performing routine SBGM, but today in my office her fasting blood sugar was 342 mg/dl.

She smokes about five cigarettes per day and does not use illicit drugs. She drinks alcohol infrequently with friends on the weekend. Her diet is generally poor with sugar-containing beverages twice daily with meals. She has no allergy to medications and denies numbness or tingling in her feet. She denies burning on urination or cloudy urine. She takes no other medication and has no known allergies to medications.

Vital signs: BP 128/65, P 80, BMI 36. Blood glucose 342 mg/dl. Acanthosis nigricans (darkened pigment indicative of insulin resistance) is present on her neck and elbows.

I make a **LIFEMAP** as follows:

WU: 6:00 a.m.

B: 6:30 a.m.

SN: 10:30 a.m.

L: 12:00 noon

SN: 2:00 p.m.

D: 5:30 p.m.

BT: 11:00 p.m.

Based on her **LIFEMAP** we construct the following rubric for blood sugar monitoring:

SBGM:

Before Breakfast (BB): 6:00 a.m.

After Breakfast (AB): 8:30 a.m.

Before Lunch (BL):

After Lunch (AL): 2:00 p.m.

Before Dinner (BD):

After Dinner (AD): 7:30 p.m.

Bedtime (BT):

We discuss that her life is very busy because she has two young children and works at a retail store four days per week from 7:30 a.m. to 5:00 p.m. She usually skips lunch when she works, but grabs snacks around 10:30 a.m. and 2:00 p.m. Her coworkers do not know she has diabetes, but she thinks she can check her blood sugar in the bathroom at 8:30 a.m. and at 2:00 p.m. We agree that she will perform SBGM two days per week; four times daily to accommodate her difficult schedule.

I do not perform a pelvic exam, but she tells me her vaginal discharge has been going on for about one month and is bothersome. She has not been sexually active for about six months. I explain to her that she will need a vaginal examination, but the most likely cause for the discharge is vaginal yeast infection due in part to uncontrolled blood sugar.

I explain to Shauna that her blood sugar control is poor, but we have a plan to control her sugar and things will improve. I explain that the **LIFEMAP** is the critical element of her care and she would be wise to bring back a glucose logbook on the next visit even if her numbers are bad. She states that nobody has ever explained diabetes to her like this before.

I tell her that she will need to take a pill once a day to protect her kidneys and I would like to start her on a dose of long-acting insulin at bedtime. She is agreeable and so I switch her to generic glargine at bedtime. I also ask her how easy it would be to give up drinking sugar-containing beverages at meals. She thinks she can do this.

I tell her that I will give her a three-day course of medicine for her presumed vaginal yeast infection, but she must follow up with a gynecologist for an exam. Before concluding I give Shauna a correction dose of lispro insulin 20 units for her fasting blood sugar of 342 mg/dl.

Visit 1 Summary:

1. Obtain diabetes relevant history.
2. Medication reconciliation.
3. Build **LIFEMAP** and SBGM at home.
4. Eliminate sugar-containing beverages with meals.
5. Begin lisinopril 5 mg daily.
6. Begin Basaglar insulin 20 units at bedtime.
7. Fluconazole 150 mg once daily x 3 days for presumed vaginal yeast infection.
8. Lispro insulin dose in office 20 units.
9. Referral for gynecology examination.
10. Will need discussion of cigarette cessation at future visit.
11. Return to Care (RTC) in 4 to 5 weeks with logbook.
12. Visit time is 60 minutes; Level 5 new patient visit coding (99205).

Case 2: Visit 2

Shauna returns after five weeks and states that her blood sugars are really high. She started taking bedtime insulin 20 units along with her two other shots of insulin, but it did not do too much. Her vaginal discharge improved after taking medication and she is having less night time urination, 1-2 times per night. She did not see a gynecologist as requested because she was busy. She brings in a logbook with representative numbers from several weeks, 4 times daily. A representative sample is shown below:

LIFEMAP WITH SBGM*

Allergies none
Cigarettes 5 per day
ETOH none

BB 6:00 a.m.	AB 8:30 a.m.	BL	AL 2:00 p.m.	BD	AD 7:30 p.m.	BT
202	269		322		210	
235	277		338		255	
195	245		288		227	
177	228		253		299	
---------	---------	----	---------	----	---------	----
---------	---------	----	---------	----	---------	----

Wake up ▸ Breakfast ▸ Snack ± ▸ Lunch ▸ Snack ± ▸ Dinner ▸ Snack ± ▸ Bedtime

Time 6:00 7:00 8:00 9:00 10:00 11:00 12:00 noon 1:00 2:00 3:00 4:00 5:00 6:00 7:00 8:00 9:00 10:00 11:00 12:00 mn

Social determinants of disease Avg. Glucose 251.3 Est. A1C 10.4%

[X] psychosocial ☐ financial ☐ access ☐ substance abuse ☐ compliance
Micro/Macrovascular Complications

☐ Neuropathy   ☐ Retinopathy   ☐ Microalbumin   ☐ Macroalbumin

☐ CVD   ☐ CVA   ☐ PVD   ☐ Carotid Disease

* **Self blood glucose monitoring**

Shauna has tried to eliminate sugar-containing beverages from her diet and has had partial success. During her visit I note that her blood sugar is highest after lunch when she does not take insulin (when she eats lunch). She states, "I tried to improve my diet in the afternoon and have eliminated soda most of the time."

Today I discuss with Shauna that she has done a great job in providing me with information we can use to improve her blood sugar control. I note that her fasting sugar is high in the morning because she needs more insulin at bedtime. I tell her we should target her fasting sugar to 90-130 mg/dl on average although not all numbers will be in that range. We agree to increase her bedtime Basaglar insulin (generic glargine) to 30 units and see if that improves her blood sugar reading at 6:00 a.m. I tell her to keep increasing her bedtime insulin by five units per week if her numbers are not falling into range. So, she can take 35 units after week 1, 40 units after week 2, and 45 units after week 3, etc. I ask her to return in four to six weeks with her logbook so we can gradually control her blood sugar.

I also increase lisinopril to 20 mg daily to afford her renal protection. She understands the instructions and agrees to go along with our plan. I check her eyes and feet today without any problems identified.

Visit 2 Summary:

1. Increase Basaglar insulin to 30 units at bedtime.
2. Continue with lispro insulin before breakfast and before dinner.
3. Increase lisinopril to 20 mg daily.
4. Continue with SBGM 2 days per week; 4 times daily.
5. RTC in 5 weeks with logbook.
6. Visit time is 25 minutes; Level 4 follow-up patient visit coding (99214).

Case 2: Visit 3

Shauna returns in five weeks and states she increased her bedtime insulin to 40 units as instructed. The last three weeks of her logbook records significant improvement in her blood sugar as shown below:

LIFEMAP WITH SBGM*

Allergies none
Cigarettes 5 per day
ETOH none

BB 6:00 a.m.	AB 8:30 a.m.	BL	AL 2:00 p.m.	BD	AD 7:30 p.m.	BT
148	185		228		193	
133	192		208		225	
118	154		194		215	
127	166		175		202	
109	148		188		190	
113	138		159		145	

Wake up · Breakfast · Snack ± · Lunch · Snack ± · Dinner · Snack ± · Bedtime

Time 6:00 7:00 8:00 9:00 10:00 11:00 12:00 noon 1:00 2:00 3:00 4:00 5:00 6:00 7:00 8:00 9:00 10:00 11:00 12:00 mn

Social determinants of disease   Avg. Glucose 168.9 Est. A1C 7.5%

[X] psychosocial ☐ financial ☐ access ☐ substance abuse ☐compliance
Micro/Macrovascular Complications

☐ Neuropathy   ☐ Retinopathy ☐ Microalbumin ☐Macroalbumin

☐ CVD   ☐ CVA   ☐ PVD   ☐ Carotid Disease

* Self blood glucose monitoring

Her fasting glucose has improved dramatically and she feels better but is concerned about weight gain. Her estimated A1C is 7.5% which is very good. Her BMI is now 38 with a 13-pound weight gain (5'6"/223 pounds to 5'6"/236 pounds). She is presently using Basaglar insulin 40 units at bedtime + lispro insulin 20 units before breakfast and dinner. We discuss that insulin causes weight gain in controlling blood sugar and that is a "nasty" consequence, but it is better to have some weight gain with improved sugar control than to have lower weight with uncontrolled blood sugar. I explain to her that we have several options:

1. A saccharidase inhibitor before each meal to decrease the absorption of carbohydrates and lessen the need for insulin (possibly).

2. Add a GLP-1 agonist to see if we can get appetite control with insulin treatment (I explain that insulin can stimulate appetite and stores fuel in lowering blood sugar, a "double whammy").

3. Consider switching her to a combination GLP-1 + insulin analog at bedtime called Soliqua.

4. Consider a bariatric surgical procedure for weight loss.

Shauna is not interested in bariatric surgery but would consider trying a saccharidase inhibitor or replacing Basaglar with a GLP-1 + insulin combination at bedtime. She asks me, "What do you think?"

I tell her the simple move is to substitute Basaglar with Soliqua (a combination of insulin glargine + lixisenatide) and see if she gets benefit from the GLP-1 agonist. I write her a prescription for Soliqua 30 units at bedtime, which contains 30 units of glargine + 10 micrograms of lixisenatide. I explain that we might need to do a prior authorization for her insurance plan to approve this new medication.

I share with her that the new drug might give her the "I'm not hungry" signal if it works. I tell Shauna that she can titrate it up to 60 units at bedtime as needed to control fasting sugar into the 90-130 range. Shauna brings the new prescription to her pharmacy and finds that it is covered with a Tier 3 copayment. She is able to afford this medication.

Visit 3 Summary:

1. Review **LIFEMAP**.
2. Discuss options for further improvement in glucose control.
3. Discontinue Basaglar and switch to Soliqua.
4. Obtain A1C, metabolic profile, complete blood count, urine micro-albumin to creatinine ratio, and lipid profile prior to next visit.
5. Continue with SBGM 2 days per week; 4 times daily.
6. RTC in 6 weeks with logbook.
7. Visit time is 25 minutes; Level 4 follow-up patient visit coding (99214).

Case 2: Visit 4

Shauna returns six weeks later and is very pleased. She states, "I am tolerating the new medication and lost five pounds. I am taking 45 units at bedtime." She notes that her blood sugar has improved significantly too. We review her LIFEMAP.

LIFEMAP WITH SBGM*

Allergies none
Cigarettes 5 per day
ETOH none

BB 6:00 a.m.	AB 8:30 a.m.	BL	AL 2:00 p.m.	BD	AD 7:30 p.m.	BT
108	144		138		168	
96	115		155		148	
125	162		177		135	
118	141		166		128	
105	172		188		157	
94	126		156		144	

Wake up ▶ Breakfast ▶ Snack ± ▶ Lunch ▶ Snack ± ▶ Dinner ▶ Snack ± ▶ Bedtime

Time 6:00 7:00 8:00 9:00 10:00 11:00 12:00 noon 1:00 2:00 3:00 4:00 5:00 6:00 7:00 8:00 9:00 10:00 11:00 12:00 mn

Social determinants of disease   Avg. Glucose 140.2 Est. A1C 6.5%

[X] psychosocial ☐ financial ☐ access ☐ substance abuse ☐ compliance
Micro/Macrovascular Complications

☐ Neuropathy ☐ Retinopathy ☐ Microalbumin ☐ Macroalbumin

☐ CVD ☐ CVA ☐ PVD ☐ Carotid Disease

* **Self blood glucose monitoring**

She has made significant progress and her A1C has improved to 7.4% with urine microalbumin of 22 and normal labs otherwise. Her LIFE-MAP estimated A1C is 6.5% reflecting lower blood sugar over the more recent weeks compared to her blood A1C level that estimates glucose over a 3-month period. She is pleased with Soliqua and feels motivated to lose more weight. She wants to continue with her present medications. I tell her she is doing great and I would like her to return in two months with her log-book. I ask her to continue monitoring her glucose two days per week, four times daily, so we can assess her progress and make sure she stays on track. I tell her we do not need additional laboratory blood tests for the next visit.

Visit 4 Summary:

1. Review LIFEMAP.
2. Continue with SBGM 2 days per week; 4 times daily.
3. Refill Soliqua.
4. RTC in 2 months with logbook.
5. Visit time is 15 minutes; Level 3 follow-up patient visit coding (99213).

Case 3
Long-standing Poorly Controlled Type 2 Diabetes with Microvascular Complications

Case 3: Visit 1

Jeff is a complex 58-year-old African-American male with history of poorly controlled type 2 diabetes for 15 years. He had laser treatment in the left eye for diabetic retinopathy about 4 years ago. He has numbness and tingling in his feet at night and takes medicine for high blood pressure. He does not smoke cigarettes or drink alcohol.

He is presently looking for better diabetes control and has been taking lispro 75/25 insulin 35 units before breakfast + 25 units before dinner and metformin (1000) mg twice daily with meals. He is taking lisinopril (10 mg) daily and metoprolol (25 mg) twice daily.

He denies significant hypoglycemia, but states that his blood sugar is high when he checks it before breakfast and before dinner. When asked about his prior diabetes care he said, "I write down my blood sugar results from home and the doctor looks at the numbers."

When asked what would happen next, Jeff said, "The doctor would tell me to keep taking my insulin and checking my blood sugar. Sometimes he would increase or decrease my insulin dose."

I told Jeff that we will do things a bit different at this visit. I asked him if he knew about hemoglobin A1C, but he said, "That's the diabetes number." When asked if he knew his most recent A1C, Jeff said, "I don't really know, but the doctor said it should be lower."

I tell Jeff that hemoglobin A1C is a number that reflects the amount of sugar that attaches to the red blood cells in his circulation. I instruct him that although it is a diabetes test it is on a different scale than blood sugar, not 100, 200, or 300. I tell him this scale runs from 5 to 20 percent and for diabetes 6 to 7 is excellent, 7 to 8 is good, 8 to 9 is fair, 9 to 10 is poor, and over 10 is uncontrolled diabetes.

I ask Jeff if he knows the names of all the medication he takes and he provides me with a list. I generate a medication reconciliation in my Electronic Medical Record. I ask Jeff if he has any medical problems that need to be addressed on this visit and he says, "No, but I would like something for my feet."

Jeff works a typical 9 a.m. to 5 p.m. job and has a regular schedule with standard mealtimes. I tell him we will build a LIFEMAP to plot out a strategy for controlling his blood sugar. Today his vital signs are as follows: BP 148/85, P 88, Ht. 70", Wt. 210; BMI 30.1.

I make a **LIFEMAP** as follows:

WU: 7:00 a.m.

B: 7:30 a.m.

L: 12:00 noon

D: 6:30 p.m.

SN: 8:00 p.m.

BT: 10:30 p.m.

Based on his **LIFEMAP** we construct the following rubric for blood sugar monitoring:

SBGM:

Before Breakfast (BB): 7:00 a.m.

After Breakfast (AB): 9:30 a.m.

Before Lunch (BL):

After Lunch (AL): 2:00 p.m.

Before Dinner (BD):

After Dinner (AD): 8:30 p.m.

Bedtime (BT):

I offer Jeff instructions to perform SBGM three days per week, four times daily, and return in four weeks with his logbook. Since he has had diabetes for many years and does not remember his last A1C, I ask him to obtain lab tests before his next visit that includes hemoglobin A1C, urine microalbumin to creatinine ratio, lipid panel, complete blood count (CBC), and comprehensive metabolic profile (CMP). I let Jeff know that he should expect poor blood sugar numbers, but not to get upset about them. We will start to build a treatment strategy on the next visit. Jeff reminds me that his feet are very numb and he would like to get medication for this problem. I give him a prescription for gabapentin 300 mg three times daily before he leaves my office. Examination of the feet reveals decreased 10-gram mono-filament sensation and hyperesthesia on vibration testing.

Visit 1 Summary:

1. Obtain diabetes relevant history.
2. Medication reconciliation.
3. Build **LIFEMAP** and SBGM at home.
4. Obtain lab tests prior to next visit.
5. Focused foot exam.

6. Gabapentin 300 mg 3 times daily as needed for neuropathy pain.
7. Return to Care (RTC) in 4 to 5 weeks with logbook.
8. Visit time is 45 minutes; Level 4 new patient visit coding (99204).

Case 3: Visit 2

Jeff returns with new laboratory data and a logbook. We look at his lab results and find an A1C of 11.5%, a urine microalbumin to creatinine ratio of 253 mcg per mg creatinine (greater than 30 mcg/mg creatinine is abnormal), cholesterol 228, triglyceride 375, HDL 34, LDL 122, BUN 25, serum creatinine 1.36, estimated glomerular filtration rate (eGFR) 69, hemoglobin 12.8, hematocrit 36.6, AST 32, ALT 28. Based on his lab numbers we can include the follow diagnoses in his problem list: chronic kidney disease stage 2, hyper-triglyceridemia (dyslipidemia), mild anemia that could be secondary to kidney disease, and possible fatty liver disease. These diagnoses are

LIFEMAP WITH SBGM*

Allergies none
Cigarettes none
ETOH none

BB 7:00 a.m.	AB 9:30 a.m.	BL	AL 2:00 p.m.	BD	AD 8:30 p.m.	BT
188	322		286		359	
221	288		326		404	
256	337		305		288	
175	268		366		332	
236	292		310		256	
178	244		328		322	

Wake up ➤ Breakfast ➤ Snack ± ➤ Lunch ➤ Snack ± ➤ Dinner ➤ Snack ± ➤ Bedtime

Time 6:00 7:00 8:00 9:00 10:00 11:00 12:00 noon 1:00 2:00 3:00 4:00 5:00 6:00 7:00 8:00 9:00 10:00 11:00 12:00 mn

Social determinants of disease   Avg. Glucose 291 Est. A1C 11.8%

☐ psychosocial ☐ financial ☐ access ☐ substance abuse ☐ compliance
Micro/Macrovascular Complications
☒ Neuropathy ☒ Retinopathy ☒ Microalbumin ☐ Macroalbumin

☐ CVD ☐ CVA ☐ PVD ☐ Carotid Disease
* **Self blood glucose monitoring**

consequences, full or in part, of poorly controlled type 2 diabetes over many years. Next, we look at Jeff's logbook using our LIFEMAP tool as shown below.

Jeff has poorly controlled blood sugar with an estimated A1C of 11.8% which approximately matches his blood A1C test. There are many places we can start his treatment plan. For simplicity, I suggest we start with bedtime insulin to control his fasting blood sugar since this is a convenient way to build a more robust insulin treatment plan.

In addition, insulin makes sense because it will improve his overall glucose control, probably decrease hepatic glucose production and insulin resistance in the liver, and lead to a better lipid profile. We agree to begin 20 units of a long-acting insulin analog at bedtime since he has a lot of room for glucose lowering.

In addition, I increase his lisinopril to 20 mg daily with the aim of maintaining him in the dosing range of 20-40 mg daily, which is the kidney protective range.[49] Jeff is eager to begin his new treatment and I suggest that his neuropathy might improve slightly due to improved blood sugar control. He has follow-up with his eye doctor already.

Visit 2 Summary:

1. Review **LIFEMAP**.
2. Add bedtime insulin 20 units.
3. Review lab results.
4. Increase lisinopril to 20 mg daily.
5. Continue with SBGM 3 days per week, 4 times daily.
6. RTC in 4-5 weeks with logbook.
7. Consider hepatology evaluation for fatty liver (future referral).
8. Medication reconciliation.
9. Visit time is 25 minutes; Level 4 follow-up visit coding (99214).

---

49   K. J. Schjoedt, A. S. Astrup, F. Persson, et al., "Optimal dose of lisinopril for renoprotection in type 1 diabetic patients with diabetic nephropathy: a randomized crossover trial," *Diabetologia*, 2009, 52:46-49.

Case 3: Visit 3

Jeff returns in five weeks and is feeling considerably better. He is pleased with his improved blood sugar control, but states that his feet are still numb. BP 132/82, P80, Wt. 212, BMI 30.4. He brings in his logbook that shows significant reduction in fasting glucose, but continued elevation of post-prandial blood sugar (this was predicted and acceptable). Jeff has tried to reduce carbohydrate content of his diet but finds it difficult. He still has trouble particularly at dinnertime when he eats rice, potatoes, or pasta as part of his meal. We discuss the use of additional insulin, but Jeff would like to try pills first. We compromise and decide to try a saccharidase inhibitor, acarbose, with the first bite of breakfast, lunch, and dinner to see if we can off-load some of the carbohydrates.

Saccharidase inhibitors inhibit the breakdown of complex carbohydrates (sugars) into mono- and disaccharides (simple sugars) in the proximal intestinal. Simple sugars can be absorbed into the blood, but

LIFEMAP WITH SBGM*

Allergies none
Cigarettes none
ETOH none

BB 7:00 a.m.	AB 9:30 a.m.	BL	AL 2:00 p.m.	BD	AD 8:30 p.m.	BT
138	225		210		252	
155	183		276		322	
127	208		263		284	
149	179		233		248	
212	286		305		328	
143	158		202		227	

Wake up → Breakfast → Snack ± → Lunch → Snack ± → Dinner → Snack ± → Bedtime

Time 6:00 7:00 8:00 9:00 10:00 11:00 12:00 noon 1:00 2:00 3:00 4:00 5:00 6:00 7:00 8:00 9:00 10:00 11:00 12:00 mn

Social determinants of disease   Avg. Glucose 221.4 Est. A1C 9.3%

☐ psychosocial ☐ financial ☐ access ☐ substance abuse ☐ compliance
Micro/Macrovascular Complications
☒ Neuropathy ☒ Retinopathy ☒ Microalbumin ☐ Macroalbumin
☐ CVD ☐ CVA ☐ PVD ☐ Carotid Disease
* **Self blood glucose monitoring**

not complex sugars. I tell Jeff to take the pill with the first bite of food and he might expect gas and bloating as a side effect. I also tell him that if he tolerates the medication and does not see any lowering of blood sugar after meals, then he can take two tablets before each meal to see if a higher dose is beneficial. I examine Jeff's feet today and he has good peripheral pulses, normal arches, and decreased vibratory and 10-gram monofilament testing indicative of diabetic neuropathy. I also send him for an annual eye exam which he did not get done.

Visit 3 Summary:

1. Review **LIFEMAP**.
2. Add acarbose 25 mg 3 times daily with meals; can increase dose to 50 mg before meals if no improvement at 25 mg.
3. Increase gabapentin to 600 mg 3 times daily for neuropathy.
4. Continue with SBGM 3 days per week, 4 times daily.
5. RTC in 4-5 weeks with logbook.
6. Ophthalmology referral reinforced.
7. Repeat A1C, CMP, CBC, lipid profile, urine M/A ratio.
8. Visit time is 25 minutes; Level 4 follow-up visit coding (99214).

Case 3: Visit 4

Jeff returns after five weeks and is feeling well. His feet have improved and he states that his blood sugar has been somewhat lower. He has been taking acarbose 50 mg two to three times daily before meals and has some mild bloating but is otherwise tolerating the medication well. He returns with new laboratory data that shows an A1C of 9.2% that is significantly better than his previous value of 11.5%. He has a repeat urine microalbumin to creatinine ratio of 180 mg per gram creatinine, cholesterol 215, triglyceride 195, HDL 38, LDL 116, BUN 22, serum creatinine 1.24, estimated glomerular filtration rate (eGFR) 77, AST 26, ALT 18. He brings in a logbook that is shown below (last 2 weeks):

LIFEMAP WITH SBGM*

Allergies none
Cigarettes none
ETOH none

BB 7:00 a.m.	AB 9:30 a.m.	BL	AL 2:00 p.m.	BD	AD 8:30 p.m.	BT
128	188		235		198	
108	156		216		223	
98	146		183		202	
133	175		191		185	
112	158		187		163	
103	144		158		176	

Wake up ▶ Breakfast ▶ Snack ± ▶ Lunch ▶ Snack ± ▶ Dinner ▶ Snack ± ▶ Bedtime ▶

Time 6:00 7:00 8:00 9:00 10:00 11:00 12:00 noon 1:00 2:00 3:00 4:00 5:00 6:00 7:00 8:00 9:00 10:00 11:00 12:00 mn

Social determinants of disease   Avg. Glucose 165.3 Est. A1C 7.4%

☐ psychosocial ☐ financial ☐ access ☐ substance abuse ☐ compliance

Micro/Macrovascular Complications

☒ Neuropathy ☒ Retinopathy ☒ Microalbumin ☐ Macroalbumin

☐ CVD ☐ CVA ☐ PVD ☐ Carotid Disease

* Self blood glucose monitoring

Blood sugar control has improved significantly, but he still has elevated post meal sugar. We discuss several options including SGLT-2 inhibitors and decide on a careful trial of SGLT-2 inhibitor with a repeat test of kidney function 2 weeks after starting the medication. Alternative strategies would be conversion to a 3-shot insulin regimen using mixed insulin at breakfast plus short-acting insulin at dinner and long-acting insulin analog at bedtime (he is already taking 3 shots of insulin daily using an unconventional treatment plan with mixed insulin before breakfast and dinner plus long-acting insulin at bedtime), conversion to a 4-shot insulin regimen with short-acting insulin before each meal, plus long-acting insulin at bedtime, or a GLP-1 analog weekly. We decide on SGLT-2 inhibitor with careful monitoring of kidney function because Jeff is not ready for a full 4-shot insulin regimen and my guess is that he has little pancreatic beta cell reserve after 15 years of diabetes, thereby making GLP-1 therapy less effective. I ask Jeff to repeat labs 2 weeks after starting SGLT-2 inhibitor and continue with SBGM. He will call me for lab results and return to care in 6 weeks.

Visit 4 Summary:

1. Review **LIFEMAP**.
2. Add SGLT-2 inhibitor at lowest dose.
3. Continue with SBGM 3 days per week, 4 times daily.
4. Obtain labs 2 weeks after starting new medication; BMP and urine microalbumin to creatinine ratio. Call me for results.
5. RTC in 6 weeks.
6. Visit time is 25 minutes; Level 4 follow-up visit coding (99214).

Case 3: Visit 5

Jeff returns six weeks later excited about his blood sugar results. His kidney function was stable with BUN 26, creatinine 1.28, and urine microalbumin to creatinine ratio 193 micrograms/gram creatinine and he has not developed urinary tract infection or genital yeast infection.

LIFEMAP WITH SBGM*

Allergies none
Cigarettes none
ETOH none

BB 7:00 a.m.	AB 9:30 a.m.	BL	AL 2:00 p.m.	BD	AD 8:30 p.m.	BT
86	132		128		147	
103	122		146		128	
97	119		135		109	
78	105		122		118	
90	115		136		148	
101	127		140		136	

Wake up · Breakfast · Snack ± · Lunch · Snack ± · Dinner · Snack ± · Bedtime

Time 6:00 7:00 8:00 9:00 10:00 11:00 12:00 noon 1:00 2:00 3:00 4:00 5:00 6:00 7:00 8:00 9:00 10:00 11:00 12:00 mn

Social determinants of disease   Avg. Glucose 119.5 Est. A1C 5.8%

☐ psychosocial ☐ financial ☐ access ☐ substance abuse ☐ compliance
Micro/Macrovascular Complications

☒ Neuropathy ☒ Retinopathy ☒ Microalbumin ☐ Macroalbumin

☐ CVD ☐ CVA ☐ PVD ☐ Carotid Disease

* Self blood glucose monitoring

His blood sugar control has improved dramatically without hypoglycemia as shown below:

He is maintained on his present medications and instructed to repeat an A1C in 8 weeks. His future A1C was 6.3% with a combination of lispro mix 75/25 insulin 35 units before breakfast + lispro mix insulin 75/25 25 units before dinner + insulin Basaglar 20 units at bedtime, acarbose 50 mg before meals, metformin 1000 mg twice daily and empagliflozin 10 mg daily. He is asked to follow up in 3 months with repeat lab tests and continue SBGM as per his **LIFEMAP**.

Case 4
Insulin-dependent Diabetes with a Basal Dilemma

Case 4: Visit 1

Jenna is a 20-year-old Caucasian female with history of GAD-65 antibody positive (immune-mediated diabetes) type 1 diabetes since age 13. She has had intermittent well-controlled blood sugar with episodes of noncompliance but no evidence of microvascular complications.

She has been on and off insulin pump therapy primarily because she is a competitive college swimmer and finds that her pump does not work well when she is swimming every day. Recently, she switched back to insulin injections because it is easier to manage.

She has tried continuous glucose monitoring but finds the device has similar problems when she is in the pool several hours a day. She is presently taking detemir insulin 20 units at bedtime plus bolus insulin before meals.

She performs loose carbohydrate counting (she eyeballs her food) and estimates her insulin dosing with an average of 6-12 units of short-acting insulin before meals. She knows about glucose monitoring but confesses that she is only doing it 3 times daily before each meal. Jenna had recent blood tests that showed an A1C of 7.0% with urine microalbumin to creatinine ratio of 22 mcg/mg creatinine.

She would like to improve her glucose control but finds she gets hypoglycemia 2-3 time per week, particularly in the morning or in the middle of the night. On exam her BP is 120/76, P 68, Ht 5'8", Wt 155 lbs., BMI 23.6. Recent ophthalmology exam 2 months ago was normal. I ask Jenna if she would consider performing glucose profiling by alternating between 3 checks per day and 5 checks per day with focus on the after lunch and bedtime blood sugar so we can see where her insulin doses need adjusting. She is willing to try this approach.

We build a **LIFEMAP** as follows:

WU: 8:00 a.m.

B: 8:30 a.m.

L: 12:30 p.m.

SN: 3:00 p.m.

D: 6:00 p.m.

SN: 8:00 p.m.

BT: 11:30 p.m.

Based on her **LIFEMAP** we construct the following rubric for blood sugar monitoring five times daily:

SBGM:

Before Breakfast (BB): 8:00 a.m.

After Breakfast (AB):

Before Lunch (BL): 12:30 p.m.

After Lunch (AL): 2:30 p.m.

Before Dinner (BD): 6:00 p.m.

After Dinner (AD):

Bedtime (BT): 11:30 p.m.

I explain to Jenna it is important for us to see what happens to her blood sugar overnight when she takes a fixed dose of long-acting insulin at bedtime. We agree she will alternate SBGM 3 checks per day before meals and 5 checks per day as shown above. We also agree she will return with her logbook in four weeks.

Visit 1 Summary:

1. Obtain diabetes-relevant history.
2. Medication reconciliation.
3. Build **LIFEMAP** and SBGM at home.
4. Return to Care (RTC) in 4 weeks with logbook.
5. Visit time is 45 minutes; Level 4 new patient visit coding (99204).

Case 4: Visit 2

Jenna returns in 4 weeks with a logbook that has alternating days with 3 glucose checks per day and 5 glucose checks per day as shown below (last week). It becomes clear that 20 units of bedtime insulin is too much when her blood sugar is lower at bedtime. We discuss the implications of this and she states previously she would not take bedtime insulin

LIFEMAP WITH SBGM*

Allergies none
Cigarettes none
ETOH none

BB 8:00 a.m.	AB	BL 12:30 p.m.	AL 2:30 p.m.	BD 6:00 p.m.	AD	BT 11:30 p.m.
127		146	133	175		210
115		133		166		
105		117	142	138		125
68		188		173		
138		147	166	182		198
56		192		163		

Wake up | Breakfast | Snack ± | Lunch | Snack ± | Dinner | Snack ± | Bedtime

Time 6:00 7:00 8:00 9:00 10:00 11:00 12:00 noon 1:00 2:00 3:00 4:00 5:00 6:00 7:00 8:00 9:00 10:00 11:00 12:00 mn

Social determinants of disease   Avg. Glucose 146 Est. A1C 6.7%

☐ psychosocial   ☐ financial   ☐ access   ☐ substance abuse   ■ compliance

Micro/Macrovascular Complications

■ Neuropathy   ☐ Retinopathy   ☐ Microalbumin   ■ Macroalbumin

☐ CVD   ☐ CVA   ☐ PVD   ■ Carotid Disease

* **Self blood glucose monitoring**

when her blood sugar was below 150 mg/dl because she feared low blood sugar overnight.

I told her this was a logical maneuver on her part, but that her blood sugar is high in the morning if she omits her bedtime insulin dose. We will fix this today. I ask Jenna if she is willing to check her blood sugar every night before bedtime and she agrees to this. Jenna always needs to check her blood sugar before meals in order to estimate her bolus insulin dose. I tell her we are not learning much from her blood sugar after lunch so we reconfigure her SBGM to BB 8:00 a.m., BL 12:30 p.m., BD 6:00 p.m., and BT 11:30 p.m. (4 times daily), with the idea that we will create a basal scale for her bedtime insulin dose based on her bedtime blood sugar. I implement a basal scale for her bedtime insulin as follows based on her bedtime glucose reading: 70-100, 6 units; 100-150, 10 units; 151-200, 15 units; 201-250, 20 units; 251-300, 25 units.

She is pleased with this approach and ready to implement her new dosing strategy. I ask her to return in 4-6 weeks so we can cross check her dosing schema. Now she will take short-acting insulin 6-12 units 3 times daily before meals + basal insulin at bedtime per scale.

Visit 2 Summary:

1. Review **LIFEMAP** with glucose values.
2. Implement basal insulin scale at bedtime.
3. Change SBGM to 4 times daily including a check at bedtime.
4. Return to Care (RTC) in 4-6 weeks with logbook.
5. Repeat A1C prior to next visit.
6. Visit time is 25 minutes; Level 4 follow-up patient visit coding (99214).

Case 4: Visit 3

Jenna returns very happy. She states her new basal scale is perfect and her fasting glucose is no longer in the hypoglycemic range. She is feeling well and thinks this will work for the long term. She has her logbook and a recent A1C that has decreased to 6.2%. Her estimated A1C per **LIFE-MAP** approximates this number at 5.8%. She has started monitoring her blood sugar every night at bedtime. I encourage Jenna to continue using her LIFEMAP to maintain excellent glucose control.

LIFEMAP WITH SBGM*

Allergies none
Cigarettes none
ETOH none

BB 8:00 a.m.	AB	BL 12:30 p.m.	AL	BD 6:00 p.m.	AD	BT 11:30 p.m.
111		128		142		153
98		116		135		118
82		105		156		122
126		111		135		95
112		115		128		86
94		123		145		137

Wake up → Breakfast → Snack ± → Lunch → Snack ± → Dinner → Snack ± → Bedtime

Time 6:00 7:00 8:00 9:00 10:00 11:00 12:00 noon 1:00 2:00 3:00 4:00 5:00 6:00 7:00 8:00 9:00 10:00 11:00 12:00 mn

Social determinants of disease   Avg. Glucose 119.7 Est. A1C 5.8%

☐ psychosocial ☐ financial ☐ access ☐ substance abuse ☐ compliance
Micro/Macrovascular Complications
☐ Neuropathy ☐ Retinopathy ☐ Microalbumin ☐ Macroalbumin
☐ CVD ☐ CVA ☐ PVD ☐ Carotid Disease
* Self blood glucose monitoring

Visit 3 Summary:

1. Review **LIFEMAP** with glucose values.
2. Continue basal insulin scale at bedtime.
3. Return to Care (RTC) in 3-4 months with logbook.
4. Repeat A1C prior to next visit.
5. Visit time is 15 minutes; Level 3 follow-up patient visit coding (99213).

Case 5
Morbid Obesity, Insulin Resistance, and Uncontrolled Type 2 Diabetes

Case 5: Visit 1

Grace is a 55-year-old Hispanic female who has suffered with obesity most of her life. She recalls being made fun of on the playground in grade school because she was big and has tried "every diet in the book" with limited success over several decades.

She is contemplating bariatric surgery for weight reduction, but her surgeon said she needs good blood sugar control prior to surgery. She has had diabetes for about 12 years and was on oral agents' glipizide plus metformin for about 5 years before being switched to insulin, liraglutide, and metformin.

Since being on insulin she had gained about 50 pounds and complains of numbness in her feet at night and intermittent pain in her knees when she walks. She is on a low carbohydrate diet and avoids simple sugars whenever possible, but states she is not perfect. She does not smoke cigarettes or drink alcohol and works at a sedentary job as a bookkeeper for a law firm.

She has not had recent blood tests but a random blood sugar today in my office is 266 mg/dl after lunch. She checks her blood sugar twice daily, before breakfast and before dinner, and states her sugar is usually in the 180-320 mg/dl range.

She denies hypoglycemia and had a normal eye exam about 6 months ago. Her blood pressure is 132/86, pulse 92, height 5'3", weight 247 pounds, and BMI 43.8. Her foot exam shows decreased vibratory sense, decreased 10-gram monofilament sensation, normal arches, no ulcers, and normal distal pulses.

Moderate acanthosis nigricans is noted on her neck and elbows. No striae or proximal muscle weakness is noted and she has no stigmata of

Cushing's syndrome (a disease of excess cortisol that can have features of obesity and diabetes.) Her present medications include liraglutide 1.8 mg daily via subcutaneous injection, metformin 1000 mg twice daily with breakfast and dinner, lisinopril 2.5 mg daily, and glargine insulin 50 units at bedtime + lispro insulin 25 units 3 times daily before meals. Total daily insulin dose is 125 units.

We discuss building a **LIFEMAP** to get a better ground level view of Grace's glucose control.

Her **LIFEMAP** is as follows:

WU: 6:30 a.m.

B: 7:30 a.m.

L: 1:00 p.m.

SN: 3:00 p.m.

D: 6:30 p.m.

BT: 10:30 p.m.

Based on her **LIFEMAP** we construct the following rubric for blood sugar monitoring 3 days per week; 4 times daily:

SBGM:

Before Breakfast (BB): 6:30 a.m.

After Breakfast (AB): 9:30 a.m.

Before Lunch (BL):

After Lunch (AL): 3:00 p.m.

Before Dinner (BD):

After Dinner (AD): 8:30 p.m.

Bedtime (BT):

Grace sleeps fitfully with episodes of nocturia and numbness in her feet. She feels fatigue frequently and thinks she has depression. She is taking a low-dose ACE inhibitor and has no treatment for diabetic neuropathy. I increase Lisinopril to 5 mg daily with intent to titrate up the medication. I start her on gabapentin 300 mg three times daily as needed for neuropathy discomfort. She will need an evaluation for sleep apnea after bariatric surgery if her sleep pattern continues to be disrupted.

Visit 1 Summary:

1. Obtain diabetes-relevant history with foot exam.
2. Medication reconciliation.
3. Build **LIFEMAP** and SBGM at home; discuss using **LIFEMAP** to get sugar under control.
4. Begin lisinopril and gabapentin.
5. Return to Care (RTC) in 4 weeks with logbook.
6. Visit time is 45 minutes; Level 4 new patient visit coding (99204).

Case 5: Visit 2

Grace returns after 5 weeks with her logbook and states her feet feel a little better, but she is still getting up frequently in the night to urinate. A sampling of her blood sugar results are shown below:

LIFEMAP WITH SBGM*

Allergies none
Cigarettes none
ETOH none

BB 6:30 a.m.	AB 9:30 a.m.	BL	AL 3:00 p.m.	BD	AD 8:30 p.m.	BT
237	322		384		301	
188	255		303		296	
228	275		315		336	
245	290		328		315	
197	248		302		285	
212	258		273		302	

Wake up ▸ Breakfast ▸ Snack ± ▸ Lunch ▸ Snack ± ▸ Dinner ▸ Snack ± ▸ Bedtime

Time 6:00 7:00 8:00 9:00 10:00 11:00 12:00 noon 1:00 2:00 3:00 4:00 5:00 6:00 7:00 8:00 9:00 10:00 11:00 12:00 mn

Social determinants of disease   Avg. Glucose 279 Est. A1C 11.3%

[X] psychosocial ☐ financial ☐ access ☐ substance abuse ☐ compliance
Micro/Macrovascular Complications
[X] Neuropathy ☐ Retinopathy ☐ Microalbumin ☐ Macroalbumin
☐ CVD ☐ CVA ☐ PVD ☐ Carotid Disease
* Self blood glucose monitoring

On review of her blood sugar pattern, Grace has fasting hyperglycemia and further post meal glucose elevation. She is already on 125 units of insulin daily and demonstrates a pattern of insulin resistance more so than insulin deficiency because her blood sugar fluctuation is modest (with insulin deficiency I would expect to see post meal glucose levels that are considerably higher or bigger swings in her blood sugar readings).

At worst she has a combination of both severe insulin resistance and insulin deficiency. Since our short-term goal is to control her glucose in preparation for bariatric surgery, I am not overly concerned about additional weight gain from insulin. We discuss several treatment possibilities including adding an SGLT-2 inhibitor, but my inclination is to switch to U-500 insulin twice daily since she will require high dose insulin to control her blood sugar (this is a special type of insulin that is 5 times concentrated). She has had full bariatric surgery evaluation and is waiting for blood sugar control. We agree to begin with Humulin R U-500 insulin with 2/3 of the dose in the morning before breakfast and 1/3 of the dose before dinner. We will start with 100 units about 40-60 minutes before breakfast and 50 units about 40-60 minutes before dinner. I ask that she check her blood sugar intermittently at 2:00 a.m. to ensure she is not developing hypoglycemia in the middle of the night. I explain to Grace that she might experience some additional weight gain, but that this is necessary as a prelude to surgery.

Visit 2 Summary:

1. Review **LIFEMAP**.
2. Discontinue 4-shot insulin regimen.
3. Begin Humulin R U-500 insulin twice daily.
4. Increase lisinopril to 10 mg daily.
5. Increase gabapentin to 600 mg three times daily.
6. Continue with SBGM 3 days per week, 4 times daily, and intermittently at 2:00 a.m.

7. Return to Care (RTC) in 6 weeks with logbook.
8. Obtain labs prior to next visit: A1C, CMP, CBC, lipid panel, urine microalbumin to creatinine ratio.
9. Visit time is 40 minutes; Level 5 established patient visit coding (99215).

Case 5: Visit 3

Grace returns after six weeks with her logbook feeling ambivalent. She says her blood sugar control has improved significantly, but she has gained 8 pounds and has swollen feet. She has also noted occasional hypoglycemia after dinner and decreased her insulin dose at dinner to 40 units after one week. She denies hypoglycemia overnight but did not record her 2:00 a.m. numbers. Her feet are less numb on gabapentin 600 mg and she is taking it twice daily now. Overall, nocturia has improved and she is sleeping through the night. Her lab tests are notable for the following: hemoglobin 13.2 g/dL, hemoglobin A1C 9.4%, BUN 20, Creat 1.15, eGFR 70, urine microalbumin to creatinine ratio 55.3, cholesterol 235, HDL 34, triglyceride 189, LDL 115.

Her physical exam shows BP 140/88, P 90, BMI 45.2, Lungs Clear, CV S1 S2 normal rate, rhythm, regular beat Extremities +1 edema in both feet, good distal pulses.

We focus her visit on several issues:
1. Blood sugar control needs a bit more improvement.
2. Fluid retention from high-dose insulin.
3. Weight gain from fluid retention.
4. Hyperlipidemia that is untreated.

Her recent **LIFEMAP** sample is shown on the next page:

I suggest to Grace that we treat three out of four problems with addition of an SGLT-2 inhibitor if she tolerates this medication. I start dapagliflozin 5 mg and recheck her kidney function in 3-4 weeks. This will lower her blood pressure, decrease fluid retention, and improve blood

LIFEMAP WITH SBGM*

Allergies none
Cigarettes none
ETOH none

BB 6:30 a.m.	AB 9:30 a.m.	BL	AL 3:00 p.m.	BD	AD 8:30 p.m.	BT
185	236		228		158	
176	215		188		169	
202	218		177		175	
193	188		205		210	
175	170		199		213	
172	183		205		218	

Wake up ▸ Breakfast ▸ Snack ± ▸ Lunch ▸ Snack ± ▸ Dinner ▸ Snack ± ▸ Bedtime

Time 6:00 7:00 8:00 9:00 10:00 11:00 12:00 noon 1:00 2:00 3:00 4:00 5:00 6:00 7:00 8:00 9:00 10:00 11:00 12:00 mn

Social determinants of disease   Avg. Glucose 194 Est. A1C 8.4%

[X] psychosocial ☐ financial ☐ access ☐ substance abuse ☐ compliance
Micro/Macrovascular Complications
[X] Neuropathy ☐ Retinopathy ☐ Microalbumin ☐ Macroalbumin
☐ CVD ☐ CVA ☐ PVD ☐ Carotid Disease
* Self blood glucose monitoring

sugar control. The side effect profile includes worsening renal function (unlikely but needs checking), vaginal yeast infections, or urinary tract infections. She will also need a statin to lower LDL cholesterol and an increased dose of lisinopril to provide renal protection.

Visit 3 Summary:

1. Review LIFEMAP.
2. Begin dapagliflozin 5 mg in the morning.
3. Continue Humulin R U-500 insulin 100 units before breakfast + 40 units before dinner.
4. Increase lisinopril to 20 mg daily.
5. Add atorvastatin 10 mg daily.
6. Continue with SBGM 3 days per week, 4 times daily, and intermittently at 2:00 a.m.
7. Check lab tests, basic metabolic profile, and urine microalbumin to creatinine ratio in 4 weeks.
8. Return to Care (RTC) in 6 weeks with logbook.

9. Visit time is 25 minutes; Level 4 established patient visit coding (99214).

Case 5: Visit 4

Grace returns in six weeks with a repeat blood and urine test at four weeks as requested. Her BUN was 18, creatinine 1.12, eGFR 72, and urine microalbumin to creatinine ratio decreased to 38. She is much happier at this visit because she has lost 8 pounds and no longer has swollen feet. Her blood pressure has also improved modestly on SGLT-2 inhibitor. Her blood sugar control is better and her bariatric surgeon is ready to perform Roux-en-Y gastric bypass surgery. Grace brings in her logbook as shown below:

LIFEMAP WITH SBGM*

Allergies none
Cigarettes none
ETOH none

BB 6:30 a.m.	AB 9:30 a.m.	BL	AL 3:00 p.m.	BD	AD 8:30 p.m.	BT
142	188		163		202	
155	166		148		187	
135	148		133		182	
176	192		158		190	
136	155		138		177	
160	172		163		210	

Wake up ▸ Breakfast ▸ Snack ± ▸ Lunch ▸ Snack ± ▸ Dinner ▸ Snack ± ▸ Bedtime

Time 6:00 7:00 8:00 9:00 10:00 11:00 12:00 noon 1:00 2:00 3:00 4:00 5:00 6:00 7:00 8:00 9:00 10:00 11:00 12:00 mn

Social determinants of disease   Avg. Glucose 165.7 Est. A1C 7.4%

[X] psychosocial ☐ financial ☐ access ☐ substance abuse ☐ compliance
Micro/Macrovascular Complications
[X] Neuropathy ☐ Retinopathy ☐ Microalbumin ☐ Macroalbumin
☐ CVD   ☐ CVA   ☐ PVD   ☐ Carotid Disease
* Self blood glucose monitoring

I tell Grace she is ready for bariatric surgery and the surgeon will request a follow-up hemoglobin A1C for confirmation of blood sugar control. As an epilogue, Grace had a follow-up hemoglobin A1C four weeks later that was reduced to 7.6% and successful bariatric surgery thereafter.

Visit 4 Summary:

1. Review LIFEMAP.
2. Continue with current medications.
3. Increase lisinopril to 40 mg daily if tolerated.
4. Continue with SBGM 3 days per week, 4 times daily, and intermittently at 2:00 a.m.
5. Repeat lab tests lipid panel, comprehensive metabolic profile, complete blood count.
6. Return to Care (RTC) after bariatric surgery.
7. Visit time is 15 minutes; Level 3 established patient visit coding (99213).

Case 6
The Complex Shift Worker

Case 6: Visit 1

Martha is a 46-year-old African-American female with history of type 2 diabetes for 6 years, obesity, hypertension, and shift working. She has no known microvascular complications of diabetes and no history of cardiovascular disease but does complain of numbness in her feet at night when she works.

She has not had recent lab tests. She has been working a night shift four days per week for the past eight years. She works on varying days, but she has four days in a row and then three days off. Her hours are 8 p.m. to 6 a.m. She does not smoke cigarettes or drink alcohol.

Martha has been struggling with her blood sugar control and weight and her primary care physician has prescribed sitagliptin-metformin 50/500 twice daily + insulin aspart 70/30 insulin 25 units twice daily although there is some confusion about the timing of her insulin dosing.

She states that when her blood sugar is too low (less than 130 mg/dl) she does not take her insulin dose, although she does not check her sugar every day before each main meal.

She does not describe hypoglycemia. Martha states that she must work at this job for the next 10 years, so there is little flexibility to change jobs. She describes her sleep pattern as fitful and she is chronically fatigued. I tell Martha she is a complex patient because of her work schedule and sleep deprivation. On her days off she tries to maintain a "regular" schedule but finds the transitions from night work to days off difficult.

I tell her that we will need to build two different **LIFEMAPs**, one for her working days and a second for her days off. Once we have constructed her two **LIFEMAPs**, then we can begin to collect data and develop a treatment strategy.

I also tell Martha she has insulin resistance and point out the darkened pigment on her elbows and knuckles called acanthosis nigricans, which is indicative of this condition. She asks me why this is, and I tell her it comes from disruption of the normal sleep-wake cycles that are important for maintaining healthy metabolism. She also has other environmental and genetic factors that affect her blood sugar and cause insulin resistance.

I ask Martha to walk me through her schedule when she works. I also ask her to take her insulin doses consistently before her morning meal and before dinner so we can see what is happening to her blood sugar with insulin. Martha eats dinner at 6:00 p.m. and then gets to work by 8 p.m.

She has a snack, usually a sandwich and piece of fruit, at 2:00 a.m., and then comes home for breakfast by 8:30 a.m. She goes to bed at 9:00 a.m. and wakes up at 2:00 p.m. She has a snack at 2:30 p.m. and then repeats her work schedule. We build **LIFEMAP 1** for Martha as follows:

D: 6:00 p.m.

SN: 2:00 a.m.

B: 8:30 a.m.

BT: 9:00 a.m.

SN: 2:30 p.m.

Based on her **LIFEMAP 1** we construct the following rubric for blood sugar monitoring:

SBGM:

Before Breakfast (BB): 8:30 a.m.

After Breakfast (AB):

Before Lunch (BL):

After Lunch (AL):

Before Dinner (BD): 6:00 p.m.

After Dinner (AD): 8:00 p.m.

After Snack (SN): 4:00 a.m. Since Martha is asleep at 10:30 a.m. we cannot get a post breakfast blood sugar reading.

Martha has a different schedule on her days off. Here, she gets home from work at 7:30 a.m. and eats breakfast by 8:30 a.m., but then does house chores and shops in the morning. She finishes her errands at around 11:30 a.m. and eats lunch at 12:00 noon. Martha takes a nap from 12:30 p.m. to 4:30 p.m. and then prepares dinner. Her dinner is at 6:00 p.m. and then she has a snack at 8:00 p.m. She goes to bed by 9:30 p.m. and typically sleeps until 4:30 a.m. She usually reads or watches TV until breakfast at 8:30 a.m. She tries to maintain the same schedule on her days off, but sometimes she has a snack at 3:00 p.m.

We build **LIFEMAP 2** for Martha as follows:

B: 8:30 a.m.

L: 12:00 noon

SN: 3:00 p.m.

D: 6:00 p.m.

SN: 8:00 p.m.

BT: 9:30 p.m.

Based on her **LIFEMAP 2** we construct the following rubric for blood sugar monitoring:

SBGM:

Before Breakfast (BB): 8:30 a.m.

After Breakfast (AB): 10:30 a.m.

Before Lunch (BL):

After Lunch (AL): 2:00 p.m.

Before Dinner (BD):

After Dinner (AD): 8:00 p.m.

Bedtime (BT):

I instruct Martha that she has a very complex schedule and we will need to obtain SBGM on her working days and days off at different times. Since her days on and off rotate every month I set up a fixed rubric for her but ask that she make a small notation of "On" or "Off" next to her SBGM so we know which days are which. I ask her to provide me with SBGM 3 days per week, 4 times daily, and then return in about 6 weeks so we can start to integrate her blood sugar data into a treatment plan.

I give her a logbook with two different plans:

LIFEMAP 1: SBGM 6:00 p.m., 8:00 p.m., 4:00 a.m., 8:30 a.m.

LIFEMAP 2: SBGM 8:30 a.m., 10:30 a.m., 2:00 p.m., 8:00 p.m.

I ask Martha to return in 6 weeks with her logbook so we can begin to build a treatment plan. I ask her to obtain baseline lab testing too.

Visit 1 Summary:

1. Obtain brief history.
2. Build **LIFEMAP 1** and **LIFEMAP 2**
3. Request SBGM 3 days per week, 4 times daily.
4. Obtain fasting HbA1C, urine microalbumin to creatinine ratio, lipid panel, CMP, CBC.
5. Visit time is 60 minutes; Level 5 new patient visit coding (99205).

Case 6: Visit 2

Martha returns six weeks later with her logbook and self-blood glucose monitoring as I prescribed. She has BP 142/86 with pulse 78, Height 5'6", Weight 208 pounds, BMI 33.6. She also had fasting lab tests that showed HbA1C 11.7%, urine microalbumin to creatinine ratio of 256 mcg per mg creatinine, BUN 18, creatinine 1.1, AST 32, ALT 28, total cholesterol 215, HDL 38, LDL 128, triglyceride 186, hemoglobin 13.2, and hematocrit 41.6. She took aspart 70/30 insulin 25 units twice daily before breakfast and dinner and did not have low blood sugar. She has been taking sitagliptin-metformin 50/500 twice daily at 8:00 a.m. and 6:00 p.m. before her main meals. **LIFEMAP 1** (working days) is shown below.

LIFEMAP1 WITH SBGM*

Allergies none
Cigarettes none
ETOH none

BB 8:30 a.m.	AB	BL.	AL	BD 6:00 p.m.	AD 8:00 p.m.	ASN 4:00 a.m.
265				248	318	327
296				218	275	303
327				267	328	402
303				222	288	256
283				197	225	268
289				244	289	305

Wake up ▸ Breakfast ▸ Snack ± ▸ Lunch ▸ Snack ± ▸ Dinner ▸ Snack ± ▸ Bedtime

Time 6:00 7:00 8:00 9:00 10:00 11:00 12:00 noon 1:00 2:00 3:00 4:00 5:00 6:00 7:00 8:00 9:00 10:00 11:00 12:00 mn

Social determinants of disease Avg. Glucose 281 Est. A1C 11.4%

[X] psychosocial ☐ financial ☐ access ☐ substance abuse [X] compliance
Micro/Macrovascular Complications
[X] Neuropathy ☐ Retinopathy [X] Microalbumin ☐ Macroalbumin
☐ CVD ☐ CVA ☐ PVD ☐ Carotid Disease

* Self blood glucose monitoring

Focusing on the blood sugar values at 6:00 p.m. and 8:00 p.m. we see Martha is under dosed with insulin or has significant insulin resistance. This notion is reinforced by a hemoglobin A1C of 11.7%, indicative of uncontrolled blood sugar. We see that 25 units of aspart 70/30 insulin about 15 minutes before her 6:00 p.m. meal does not lower her blood sugar 2 hours later at 8:00 p.m. and therefore, one option for Martha is to increase her insulin dosing before meals.

Additionally, there is little variance in her blood sugar numbers when looking across rows of **LIFEMAP 1** (e.g., blood sugars range from 267 to 402 mg/dl one day and 244 to 305 mg/dl on another day). A patient with full insulin-dependent diabetes and good insulin sensitivity might show blood sugar ranging from 67 to 350 mg/dl in comparison.

This suggests that hepatic glucose production and insulin resistance are key factors in Martha's hyperglycemia more so than beta cell insufficiency, since we would expect big fluctuations in her blood sugars reading before and after meals if she was sensitive to insulin and had prominent beta cell deficiency. Either way, addition of an SGLT-2 inhibitor would potentially be useful for Martha if it did not interfere with her kidney function or cause urinary tract or genital yeast infections.

We discuss these points and decide to increase insulin dosing before meals to 35 units twice daily since Martha is concerned about excessive urination from SGTL-2 inhibition during her job. We consider the possibility that SGLT-2 might be useful on her off days when excess urination is not as much a concern.

We also recognize that sitagliptin is probably not effective in our treatment plan, but Martha has a lot of sitagliptin-metformin at home and we decide not to change this medication right now (in the future we will switch her to Metformin only).

Next, we analyze **LIFEMAP 2** as shown on the next page:

Analysis of **LIFEMAP 2** is not appreciably different from **LIFEMAP 1**. We discuss possible treatment strategies and Martha is willing to try an SGLT-2 inhibitor on her off days to compare the difference between these two treatment strategies. I am supportive of this decision with the caveat that we will need to test her urine microalbumin to creatinine ratio and serum BUN and creatinine about 2 to 3 weeks after starting SGLT-2 inhibitor to ensure it is not causing kidney dysfunction.

LIFEMAP2 WITH SBGM*

Allergies none
Cigarettes none
ETOH none

BB 8:30 a.m.	AB 10:30 a.m.	BL	AL 2:00 p.m.	BD	AD 8:00 p.m.	BT
237	285		308		286	
215	246		263		294	
196	244		228		257	
188	238		267		302	
223	275		248		285	
178	196		224		248	

Wake up — Breakfast — Snack ± — Lunch — Snack ± — Dinner — Snack ± — Bedtime

Time 6:00 7:00 8:00 9:00 10:00 11:00 12:00 noon 1:00 2:00 3:00 4:00 5:00 6:00 7:00 8:00 9:00 10:00 11:00 12:00 mn

Social determinants of disease   Avg. Glucose 247 Est. A1C 10.2%

[X] psychosocial [ ] financial [ ] access [ ] substance abuse [X] compliance

Micro/Macrovascular Complications

[X] Neuropathy [ ] Retinopathy [X] Microalbumin [ ] Macroalbumin

[ ] CVD [ ] CVA [ ] PVD [ ] Carotid Disease

* Self blood glucose monitoring

Visit 2 Summary:

1. Interrogate **LIFEMAP1** and **LIFEMAP2**.
2. Increase aspart 70/30 insulin to 35 units twice daily on working days.
3. Maintain aspart 70/30 insulin at 25 units twice daily on off days.
4. Begin empagliflozin 10 mg in the morning on off days; continue with sitagliptin-metformin and aspart 25 units twice daily.
5. Obtain urine microalbumin to creatinine ratio and CMP two week after starting empagliflozin.
6. Begin lisinopril 5 mg daily with dose titration in future.
7. Referral to Ophthalmology.
8. Return to care with **LIFEMAP1** and **LIFEMAP2** in 6 weeks.
9. Visit time is 45 minutes; Level 5 established patient visit coding (99215).

Case 6: Visit 3

Martha returns after six weeks with her logbook. She has gained two pounds but feels well. Two weeks after starting SGLT-2 inhibitor she had stable kidney function with a slight reduction in urine microalbumin to creatinine ratio to 178 mcg/mg creatinine. She says, "The increased urination does not bother me too much." She has noticed some reduction in her glucose levels, particularly on her days off. Increasing insulin dose on the last visit also seems to have helped her a little. **LIFEMAP1** is shown below. Interrogation of **LIFEMAP1** reveals a decrease in average blood glucose from her preliminary **LIFEMAP1** of 281 mg/dl to 245 mg/dl. She denies hypoglycemia and has tried to limit carbohydrates in her diet.

LIFEMAP1 WITH SBGM*

Allergies none
Cigarettes none
ETOH none

BB 8:30 a.m.	AB	BL	AL	BD 6:00 p.m.	AD 8:00 p.m.	ASN 4:00 a.m.
268				205	238	258
227				196	218	249
266				222	270	283
225				189	226	248
278				187	244	299
256				227	302	289

Wake up ▸ Breakfast ▸ Snack ± ▸ Lunch ▸ Snack ± ▸ Dinner ▸ Snack ± ▸ Bedtime ▸

Time 6:00 7:00 8:00 9:00 10:00 11:00 12:00 noon 1:00 2:00 3:00 4:00 5:00 6:00 7:00 8:00 9:00 10:00 11:00 12:00 mn

Social determinants of disease   Avg. Glucose 244.6 Est. A1C 10.1%

[X] psychosocial [ ] financial [ ] access [ ] substance abuse [X] compliance
Micro/Macrovascular Complications
[X] Neuropathy [ ] Retinopathy [X] Microalbumin [ ] Macroalbumin
[ ] CVD   [ ] CVA   [ ] PVD   [ ] Carotid Disease
* Self blood glucose monitoring

However, she has noticed a more significant decrease in SBGM when she takes SGLT-2 inhibitor. She is concerned about frequent urination at work and does not think she can take this medication on those days. **LIFEMAP 2** is show below:

LIFEMAP2 WITH SBGM*

Allergies none
Cigarettes none
ETOH none

BB 8:30 a.m.	AB 10:30 a.m.	BL	AL 2:00 p.m.	BD	AD 8:00 p.m.	BT
188	232		226		197	
175	210		238		244	
169	193		189		212	
215	249		233		208	
187	184		202		195	
178	223		202		229	

Wake up Breakfast Snack ± Lunch Snack ± Dinner Snack ± Bedtime

Time 6:00 7:00 8:00 9:00 10:00 11:00 12:00 noon 1:00 2:00 3:00 4:00 5:00 6:00 7:00 8:00 9:00 10:00 11:00 12:00 mn

Social determinants of disease   Avg. Glucose 207.4 Est. A1C 8.9%

[X] psychosocial [ ] financial [ ] access [ ] substance abuse [X] compliance
Micro/Macrovascular Complications
[X] Neuropathy [ ] Retinopathy [X] Microalbumin [ ] Macroalbumin
[ ] CVD [ ] CVA [ ] PVD [ ] Carotid Disease
* Self blood glucose monitoring

Martha's average blood sugar decreased from 245 mg/dl to 207 mg/dl on Empagliflozin. Based on her overall profile I recommend Martha increase Novolog 70/30 insulin to 45 units twice daily on working days and 35 units twice daily on off days.

I also recommend discontinuing sitagliptin-metformin and converting her to metformin 1000 mg twice daily only. Sitagliptin, like all (dipeptidyl peptidase IV) DPP-IV inhibitors that indirectly potentiate glucose-stimulated insulin secretion has weak A1C lower effects when patients are already on twice-daily insulin dosing. We can also optimize her SGLT-2 inhibitor by increasing the dose of empagliflozin to 25 mg daily. Her ophthalmology exam showed mild background diabetic retinopathy bilaterally.

Visit 3 Summary:

1. Interrogate **LIFEMAP1** and **LIFEMAP2**.
2. Increase aspart 70/30 insulin to 45 units twice daily on working days.

3. Increase empagliflozin to 25 mg in the morning on off days.

4. Discontinue sitagliptin-metformin and switch to metformin 1000 mg twice daily.

5. Obtain Hemoglobin A1C, CMP, CBC, lipid panel, urine micro-albumin to creatinine ratio prior to next visit.

6. Increase lisinopril to 10 mg daily.

7. Return to care with **LIFEMAP1** and **LIFEMAP2** in 6 weeks.

8. Repeat A1C and urine microalbumin to creatinine ratio prior to next visit.

9. Visit time is 45 minutes; Level 5 established patient visit coding (99215).

Case 6; Visit 4

Martha returns to care in 6 weeks with a smile on her face. She has noticed significant improvement in her blood sugar readings, her recent A1C has decreased to 8.4%, and her urine microalbumin to creatinine ratio is now 84 mcg/mg. She denies hypoglycemia, urinary tract, or yeast infections, and she has noticed better energy. She comes in with her **LIFEMAP 1 and 2** as shown on the next page:

Her blood sugar control has continued to improve and we discuss her dietary choices. Her average glucose has shown further lowering on **LIFEMAP 1** from 245 mg/dl to 198 mg/dl. This would translate to an estimated A1C of 8.5%. **LIFEMAP 2** has also continued to improve with higher dose empagliflozin as shown below:

Average blood sugar for **LIFEMAP2** is now 185 mg/dl. Martha is now on track toward good blood sugar control. She has not developed episodes of low blood sugar so we decide to increase aspart 70/30 insulin to 55 units before meals and maintain her other medications. We agree that a good short-term goal would be to lower A1C to <8.0% which would be equal to an average glucose level of 183 mg/dl. This would significantly decrease (but not eliminate) further risk of complications from diabetes.

LIFEMAP1 WITH SBGM*

Allergies none
Cigarettes none
ETOH none

BB 8:30 a.m.	AB	BL	AL	BD 6:00 p.m.	AD 8:00 p.m.	ASN 4:00 a.m.
185				178	205	237
207				192	238	224
195				165	188	218
210				173	180	168
193				223	238	202
177				181	175	194

Wake up → Breakfast → Snack ± → Lunch → Snack ± → Dinner → Snack ± → Bedtime

Time 6:00 7:00 8:00 9:00 10:00 11:00 12:00 noon 1:00 2:00 3:00 4:00 5:00 6:00 7:00 8:00 9:00 10:00 11:00 12:00 mn

Social determinants of disease   Avg. Glucose 197.8 Est. A1C 8.5%

[X] psychosocial   [ ] financial   [ ] access   [ ] substance abuse   [ ] compliance

Micro/Macrovascular Complications

[X] Neuropathy   [ ] Retinopathy   [X] Microalbumin   [ ] Macroalbumin

[ ] CVD   [ ] CVA   [ ] PVD   [ ] Carotid Disease

* Self blood glucose monitoring

LIFEMAP2 WITH SBGM*

Allergies none
Cigarettes none
ETOH none

BB 8:30 a.m.	AB 10:30 a.m.	BL	AL 2:00 p.m.	BD	AD 8:00 p.m.	BT
158	163		188		195	
177	188		215		202	
185	143		179		173	
199	198		223		216	
149	155		172		168	
183	189		215		196	

Wake up → Breakfast → Snack ± → Lunch → Snack ± → Dinner → Snack ± → Bedtime

Time 6:00 7:00 8:00 9:00 10:00 11:00 12:00 noon 1:00 2:00 3:00 4:00 5:00 6:00 7:00 8:00 9:00 10:00 11:00 12:00 mn

Social determinants of disease   Avg. Glucose 184.5 Est. A1C 8.1%

[X] psychosocial   [ ] financial   [ ] access   [ ] substance abuse   [ ] compliance

Micro/Macrovascular Complications

[X] Neuropathy   [ ] Retinopathy   [X] Microalbumin   [ ] Macroalbumin

[ ] CVD   [ ] CVA   [ ] PVD   [ ] Carotid Disease

* Self blood glucose monitoring

Visit 4 Summary:

1. Interrogate **LIFEMAP 1** and **LIFEMAP 2**.
2. Increase aspart 70/30 insulin to 55 units twice daily on working days.
3. Increase lisinopril to 20 mg daily.
4. Return to care with **LIFEMAP 1** and **LIFEMAP 2** in 8 weeks.
5. Repeat A1C and urine microalbumin to creatinine ratio prior to next visit.
6. Visit time is 30 minutes; Level 4 established patient visit coding (99214).

Case 6; Visit 5

Martha returns with her **LIFEMAPs** and new laboratory tests. She is smiling because her A1C has improved to 7.7% and her urine micro-albumin to creatinine ratio is now 42 mcg/mg, just slightly above the upper limits of normal (30 mcg/mg). She is feeling well and states she is amazed that she achieved her goals. We decide her insulin dose is probably maximum because she had one episode of relative hypogly-cemia when her blood sugar went down to 106 mg/dl. I explain, "This can happen when your body is used to having blood sugars up in the 200-300 range and the sugar is rapidly lowered to the low 100s." She states she drank some juice and her sweating and jitteriness resolved in 10 minutes.

We decide to keep her medications the same and have her continue with SBGM; however, I offer over to Martha the possibility of using a continuous glucose monitoring device so she can have blood sugar read-ings whenever she wants without having to prick her finger. I caution that this is contingent on her insurance covering the cost of the device. She is excited to try this newer technology because she is getting tired of constantly pricking her finger for a drop of blood. I prescribe a continu-ous glucose-monitoring device for Martha and tell her she can check her blood sugar any time, but to also stick with the **LIFEMAPs** that have been so successful. She agrees.

I ask her to return in three months with new blood and urine tests. I explain that she can easily fall off the railroad tracks with blood sugar control, but the **LIFEMAP** will tell her beforehand if this is happening. She understands that lifelong glucose monitoring will be an integral part of her health and well-being.

Visit 4 Summary:

1. Maintain aspart 70/30 insulin 55 units twice daily on working days and 35 units twice daily on off days.
2. Prescribe continuous glucose monitoring device.
3. Return to care with **LIFEMAP1** and **LIFEMAP2** in 3 months.
4. Repeat A1C and urine microalbumin to creatinine ratio prior to next visit.
5. Visit time is 30 minutes; Level 4 established patient visit coding (99214).

Case 7
Noncompliance with Psychosocial Stressors

Case 7: Visit 1

Patricia is a 48-year-old Caucasian female with history of type 2 diabetes for 3 years, hypertension, hyperlipidemia, and obesity with BMI 34. She has a current 10-year history of moderate alcohol consumption, estimating her weekly intake to 2-3 bottles of white wine. She was a former cigarette smoker, 1/2 pack per day for 10 years, and quit about 10 years ago.

She is presently taking metformin (1000 mg) twice daily, glipizide 10 mg daily, and liraglutide 1.8 mg daily. Her recent A1C was 9.3%. She tries to be careful with carbohydrate intake but finds that she gets hungry frequently and cannot control her appetite. She has a lot of stress in her life but does not elaborate.

She is peri-menopausal and has vaginal dryness, but no hot flashes. She notes weight gain of ~20 pounds since her menses started to decrease. She walks her dog daily for 20 minutes twice a day and gets little formal exercise. On intake her BP is 136/86, P 88, Ht 5'4", Wt 190 pounds, BMI 32.6. She denies numbness, hypoglycemia, or kidney problems and had a recent eye exam that was normal.

She is interested in improving her blood sugar control but works long hours at a supermarket to support her two college-age children. She has a OneTouch glucometer at home with adequate numbers of lancets and test strips. She is not performing routine blood sugar testing. She is a single mom who has to "do it all."

We make a **LIFEMAP** for Patricia:

WU: 6:30 a.m.

B: 7:00 a.m.

SN: 10:30 a.m.

L: 2:00 p.m.

D: 6:30 p.m.

SN: 9:00 p.m.

BT: 10:30 p.m.

Based on her LIFEMAP we decide on self-blood glucose monitoring at the following times:

SBGM:

Before Breakfast (BB): 6:30 a.m.

After Breakfast (AB): 9:00 a.m.

Before Lunch (BL):

After Lunch (AL): 4:00 p.m.

Before Dinner (BD):

After Dinner (AD): 8:30 p.m.

Bedtime (BT):

I explain to Patricia that she will need to obtain SBGM 2-3 days per week, 4 times daily, and return with information recorded in a logbook in about 4 to 5 weeks. Patricia seems willing to participate in her care. She has some concern about being able to check her blood sugar at work. I told her to try taking bathroom breaks at the designated times. I reinforced that she only needs to do this once or twice during the week and can use a weekend day for her third day.

Medication List:

✓ glipizide ER 10 mg daily

✓ metformin 1000 mg twice daily with breakfast and dinner

✓ liraglutide 1.8 mg daily

✓ simvastatin 20 mg daily at bedtime

✓ lisinopril 10 mg daily

Visit 1 Summary:

1. Build **LIFEMAP**.

2. Obtain SBGM 3 days per week, 4 times daily.

3. Medication reconciliation with refills.

4. Return to care in 4-5 weeks with glucose logbook.

5. Visit time is 45 minutes; Level 4 new patient visit coding (99204).

Case 7: Visit 2

Patricia returns in 5 weeks for her second visit. She was rushed and forgot to bring her logbook. She is upset because of family issues she does not want to describe in detail. She states that she has had a lot of stress over the past month with her mother and boyfriend. She also states that she has been stress eating, and today in my office her random blood sugar was 248 mg/dl and her weight was 194 pounds. She is feeling a bit depressed at not following through with my previous recommendations and thinks it will be hard for her to do so in the future. She states, "Why don't you just give me pills to control my sugar like they show on TV?"

I tell Patricia that managing diabetes is more than pills and involves commitment to small tweaks in lifestyle and blood sugar monitoring. Pills can be useful, but sometimes the trajectory of the disease requires shots to control the blood sugar. She emphatically states that she will not use insulin because she has seen what it does to her diabetic friend.

LIFEMAP WITH SBGM*

Allergies none
Cigarettes none
ETOH 2 bot/week

BB 6:30 a.m.	AB 9:00 a.m.	BL	AL 4:00 p.m.	BD	AD 8:30 p.m.	BT
----------------	----------------		----------------		----------------	
----------------	----------------		----------------		----------------	
----------------	----------------		----------------		----------------	
----------------	----------------		----------------		----------------	
----------------	----------------		----------------		----------------	
----------------	----------------		----------------		----------------	

Wake up › Breakfast › Snack ± › Lunch › Snack ± › Dinner › Snack ± › Bedtime

Time 6:00 7:00 8:00 9:00 10:00 11:00 12:00 noon 1:00 2:00 3:00 4:00 5:00 6:00 7:00 8:00 9:00 10:00 11:00 12:00 mn

Social determinants of disease Avg. Glucose Est. A1C

[X] psychosocial [ ] financial [ ] access [ ] substance abuse [X] compliance
Micro/Macrovascular Complications
[ ] Neuropathy [ ] Retinopathy [ ] Microalbumin [ ] Macroalbumin
[ ] CVD [ ] CVA [ ] PVD [ ] Carotid Disease
* Self blood glucose monitoring

I see the visit spiraling out of control and recognize that we need to change gears. I suggest to Patricia that she seems stressed and maybe a bit depressed. She nods in agreement. I tell her that we can use a new pill to help her, but we need to get blood sugar readings at home since that is where her diabetes care resides. She expresses relief that she will not need insulin shots today, but I emphasize that much of what happens in the future for controlling her blood sugar will depend on her actions now. I explain that I am giving her a medicine that will cause sugar to go out in the urine but will also lower her blood pressure. She will need to monitor her blood pressure and be aware that this medication called canagliflozin can cause urine infections and vaginal yeast infections in a small percent of individuals. I also explain to Patricia that we can insert a device in the office that will automatically take blood sugar readings over the next 14 days so we can obtain information at home. She will see the diabetes care coordinator directly after me to have a Freestyle Libre Pro placed on the back of her arm. This will allow us to see her SBGM every 5 minutes at the next visit. She is agreeable to this plan and states that she appreciates my working with her and not against her. I reinforce that she should drink plenty of water and try to focus effort on her own well-being. I ask if she thinks psychological support would be helpful. She states, "Not right now, but maybe in the future." I ask if she can return to my office again in 2-3 weeks. She says she will schedule a follow-up appointment. She has no acute complaints at this visit, so I do not perform a physical exam. I explain that she will need a lab test about 2-3 weeks after starting this new medication to ensure it does not affect the kidneys adversely.

Visit 2 Summary:

1. Begin canagliflozin 100 mg daily in the morning.
2. Place Freestyle Libre Pro to obtain SBGM and transition off **LIFEMAP** to continuous glucose monitoring.
3. Obtain basic metabolic profile and urine microalbumin to creatinine ratio about 2-3 weeks after starting canagliflozin.
4. Try to limit carbohydrates in diet and avoid stress eating.
5. Discuss psychological support again at next visit.
6. Return to care in 2-3 weeks.
7. Visit time is 25 minutes; Level 4 established patient visit coding (99214) plus technical component for placement of Freestyle Libre Pro by physician and office staff. (95250)

Case 7: Visit 3

Patricia returns 3 weeks later. We download information from her Freestyle Libre Pro to obtain continuous glucose monitoring with particular focus on extraction of data points before breakfast and 2 hours after meals.

Today in my office Patricia seems a bit brighter and her random glucose is 183 mg/dl. Canagliflozin has worked nicely to improve her blood sugar readings. They have ranged from 98-153 in the morning, 137-228 ~2 hours after breakfast, 142-218 ~2 hours after lunch, and 156-245 ~ 2 hours after dinner. She has lost 5 pounds and weighs 189 with BP 112/ 72. Her kidney function is stable and she does not have urogenital complaints.

LIFEMAP WITH SBGM*

Allergies none
Cigarettes none
ETOH 2 bot/week

BB 6:30 a.m.	AB 9:00 a.m.	BL	AL 4:00 p.m.	BD	AD 8:30 p.m.	BT
98-153	137-228		142-218		156-245	

Wake up ▶ Breakfast ▶ Snack ± ▶ Lunch ▶ Snack ± ▶ Dinner ▶ Snack ± ▶ Bedtime

Time 6:00 7:00 8:00 9:00 10:00 11:00 12:00 noon 1:00 2:00 3:00 4:00 5:00 6:00 7:00 8:00 9:00 10:00 11:00 12:00 mn

Social determinants of disease   Avg. Glucose   Est. A1C

[X] psychosocial [ ] financial [ ] access [ ] substance abuse [X] compliance

Micro/Macrovascular Complications

[ ] Neuropathy [ ] Retinopathy [ ] Microalbumin [ ] Macroalbumin

[ ] CVD [ ] CVA [ ] PVD [ ] Carotid Disease

* Self blood glucose monitoring

She states her blood sugar is probably high after meals when she has too many carbs. I ask if she drinks any sugar-containing beverages with meals and states, "Only tea." I ask if it is iced tea with sugar and she says, "I'll have to check, but I think yes."

I tell her this is great because it is "low-hanging fruit" to improve her post meal glucose levels. She looks at me and says, "I think I can drink something else without sugar." I tell her, "That will make a major impact on your blood sugar and we will probably be at target with that tweak."

Patricia is pleased and I say, "Let's do some blood tests in 5-6 weeks after you change your beverage with meals." Patricia looks at me quizzically and says, "What about my food?" I tell her, "It all comes out in the wash with your blood sugar readings and weight. We don't need to reinvent your diet right now, especially if your blood sugar readings are showing improvement. If you see elevated blood sugar consistently after lunch and dinner, then I am prescribing you a pill called acarbose that you can take with the first bite of food. This pill will limit carbohydrate absorption in your GI tract and lower your blood sugar two hours after eating. You can decide if you need it."

Visit 3 Summary:

1. Download Freestyle Libre Pro CGM data and interpret information.
2. Eliminate sugar-containing beverages from meals.
3. Prescribe acarbose 25 mg before lunch and dinner as needed.
4. SBGM if possible after CGM.
5. Obtain Hemoglobin A1C, lipid panel, CBC, and CMP.
6. RTC in 6 weeks.
7. Visit time is 25 minutes; Level 4 established patient visit coding (99214) plus analysis and interpretation of Freestyle Libre Pro glucose data. (95251).

Case 7: Visit 4

Patricia returns today with a smile on her face. She has successfully eliminated iced tea with meals and she brings in her partially complete glucose logbook showing after-meal glucose levels between 138-178 mg/dl. She decided to prick her finger after all. She has lost 3 pounds more and feels more energetic with confidence that her new diabetes treatment is on track. I reinforce her confidence by telling her that her HbA1C has improved to 7.5% and the rest of her numbers are stable, including kidney and liver function. Her BP today is 108/68 and pulse 86.

She recognizes that it has been hard work to get her diet straightened out and eliminate sugar-containing beverages, but she is happy with her renewed level of energy and stamina. I tell Patricia that diabetes is not a binary disease; it has a progressive trajectory if not attended to. She understands and I reinforce that home glucose monitoring will give us a heads-up if she starts to "veer off the railroad tracks."

She understands what I am saying. I ask if she thinks she can continue to perform SBGM and she agrees to try. I tell her to pick 1-2 days per week and check SBGM 4 times daily as previously recommended in her LIFEMAP. I exam her feet today and her vibration sense and 10-gram monofilament testing are normal. Her pulses are normal in the feet bilaterally. I suggest to Patricia that we get back together in 3 months with a new set of lab tests, but reemphasize that the blood test will only be confirmatory of her SBGM at home (i.e., we should have no surprises with her future A1C level. I tell her our goal is to get her A1C down to ~7.0% or lower. She agrees. I increase canagliflozin to 300 mg daily since it will give her additional lowering of A1C.

Visit 4 Summary:

1. Review laboratory data.
2. Reconcile and renew medications.
3. Increase canagliflozin to 300 mg daily
4. Repeat lab tests in 10 weeks.
5. RTC in 3 months with logbook.
6. Visit time is 15 minutes; Level 3 established patient visit coding (99213).

Case 8
Dialysis Complicating Type 2 Diabetes

Case 8: Visit 1

Harry is a 58-year-old Asian male with a long-standing history of poorly controlled type 2 diabetes for 20 years, hypertension, and neuropathy in both feet. He has been receiving hemodialysis for about 2 years and complains of both hyperglycemia and hypoglycemia. He is on a kidney transplantation list and has been using a continuous glucose monitoring device, but he feels that despite knowing his blood sugar readings he still has frequent blood sugars as low as 38 mg/dl and then up to 450 mg/dl.

He does not feel his blood sugar when it goes low, so that is frightening to him. He denies alcohol or cigarette use. He is taking 4 shots of insulin daily, using glargine insulin at bedtime 25 units plus lispro insulin 10-15 units before meals. He is also taking lisinopril (20 mg) daily for blood pressure control, atorvastatin (40 mg) daily, a baby aspirin 81 mg at bedtime, and calcium phosphate tablets (667 mg) 3 times daily before each meal.

He was offered an insulin pump in the past, but did not want to wear a device at that time. He feels like the continuous glucose monitor provides him with a lot of information, but he is not sure how to use it. I ask Harry if he has ever been given two separate insulin dosing schedules and he looks puzzled. I continue, "What days of the week do you have dialysis?"

He tells me, "Monday, Wednesday, and Friday." I say, "We will build two **LIFEMAPs** for you and see what can be done to improve your sugar control." Harry is intrigued, but still confused. I explain, "You might have two entirely different schedules on your dialysis days and non-dialysis days, so it is not logical to use the same insulin dosing strategy on these different days." His eyes light up a bit. "Let's start," I say. "We will begin with your dialysis days. What time do you get up in the morning for dialysis?" Harry tells me 5:00 a.m. He then says, "I have to be at the dialysis center by 7:00 a.m., so it is an early morning."

I ask, "Do you eat breakfast before you have dialysis or do you have food at the center?"

He says, "I typically get a bagel with cream cheese and black coffee on my way to the dialysis center and eat during dialysis around 7:30 a.m."

I continue, "Do you get your blood sugar checked at the dialysis center before you eat?"

Harry says, "Yes, if my blood sugar is above 100 mg/dl; then I take my insulin about 10 units and eat my bagel."

I continue, "What time do you finish dialysis?"

Harry says, "About 12:00 noon."

I say, "Do you eat lunch after that?"

Harry replies, "Yes, I have lunch when I get home around 1:00 p.m."

"What happens next?" I ask.

Harry says, "I usually take a nap for 2-3 hours and then go for a walk if the weather is okay."

"What time is dinner, Harry?"

"About 6:30 p.m."

"And bedtime?"

"About 10:00 p.m."

"Okay," I say, "I'm not going to get into details about food and snacks at this visit, but I would also like to get your schedule for non-dialysis days. What time do you get up on non-dialysis days? Run me through your typical day like we just did."

Harry tells me, "Usually I get up around 7:30 a.m., breakfast at 8:00 a.m., lunch at 1:00 p.m., dinner at 6:30 p.m. and bedtime at 10:30 p.m."

"What do you eat for breakfast?" I ask. Harry tells me it is usually a bowl of oatmeal without sugar or honey and a cup of black coffee. I ask,

"Do you know if there is a difference in your blood sugar after breakfast with these two different breakfast meals?" He doesn't know. I say, "Well, let's find out."

Based on his **LIFEMAP** we decide on self-blood glucose monitoring at the following times:

LIFEMAP for Dialysis Days

SBGM:

Before Breakfast (BB): 5:00 a.m.

After Breakfast (AB): 9:30 a.m.

Before Lunch (BL):

After Lunch (AL): 3:00 p.m.

Before Dinner (BD):

After Dinner (AD): 8:30 p.m.

Bedtime (BT):

LIFEMAP for Non-Dialysis Days

SBGM:

Before Breakfast (BB): 7:30 a.m.

After Breakfast (AB): 10:00 a.m.

Before Lunch (BL):

After Lunch (AL): 3:00 p.m.

Before Dinner (BD):

After Dinner (AD): 8:30 p.m.

Bedtime (BT):

I check Harry's feet before he leaves the office and find absent vibration sensation, absent 10-gram monofilament testing, and mild warmth in the right foot with tenderness in the right arch when I apply firm pressure.

I have concern that he might be developing Charcot's arthropathy (neuropathic bone disease of the foot) and order bilateral X-rays of both feet and ankles. I ask Harry to return in 4 to 5 weeks with a logbook using the two separate **LIFEMAPs** we created and results of his X-ray exam. I tell him not to change his insulin dosing at present until we collect several datasets.

Medication List:
- ✓ glargine insulin 25 units at bedtime
- ✓ lispro insulin 10-15 units 3 times daily before meals
- ✓ aspirin 81 mg daily
- ✓ atorvastatin 40 mg daily at bedtime
- ✓ lisinopril 20 mg daily
- ✓ calcium phosphate 667 mg 3 times daily with meals

Visit 1 Summary:
1. Build **LIFEMAPs**.
2. Medication reconciliation.
3. Obtain SBGM every day with different times, 4 times daily.
4. Obtain bilateral X-rays of feet and ankles to assess for Charcot's arthropathy.
5. Return to care in 4-5 weeks with glucose logbook and X-ray results.
6. Visit time is 60 minutes; Level 5 New Patient visit coding (99205).

Case 8: Visit 2

Harry returns after five weeks with two glucose logbooks, one for dialysis days and one for non-dialysis days. We formulate his datasets based on a sampling of his glucose results in two discrete **LIFEMAPS**. We begin with the **LIFEMAP** on dialysis days as shown on next page:

I begin with Harry's **LIFEMAP** on dialysis days. I ask him, "What happens on dialysis mornings when your blood sugar is below 100?"

He responds, "I usually don't take insulin because I'm afraid of going low during dialysis."

I say to him, "It appears you need insulin even when your blood sugar is <100, but perhaps 10-15 units is too much?" I then probe, "What about when you blood sugar is 108 or 147 before breakfast?"

LIFEMAP WITH SBGM*

Dialysis Days

Allergies none
Cigarettes none
ETOH none

BB 5:00 a.m.	AB 9:30 a.m.	BL	AL 3:00 p.m.	BD	AD 8:30 p.m.	BT
108	235		242		174	
95	310		264		205	
147	197		202		158	
68	331		255		229	
128	185		174		148	
83	284		296		303	

Wake up ▸ Breakfast ▸ Snack ± ▸ Lunch ▸ Snack ± ▸ Dinner ▸ Snack ± ▸ Bedtime

Time 6:00 7:00 8:00 9:00 10:00 11:00 12:00 noon 1:00 2:00 3:00 4:00 5:00 6:00 7:00 8:00 9:00 10:00 11:00 12:00 mn

Social determinants of disease   Avg. Glucose 200.9 Est. A1C 8.6%

☐ psychosocial ☐ financial ☐ access ☐ substance abuse ☐ compliance

Micro/Macrovascular Complications

☒ Neuropathy ☐ Retinopathy ☒ Dialysis ☐ Macroalbumin

☐ CVD ☐ CVA ☐ PVD ☐ Carotid Disease

* Self blood glucose monitoring

Harry replies, "I take 10 units for 108 and 15 units for 147."

I reply, "Those doses of insulin look pretty good, but look at your blood sugar results 2 hours after breakfast when your sugar before breakfast is <100. It's quite elevated! This tells us you need a dose of insulin possibly 5-10 units even when you sugar starts off low before breakfast, especially if you eat a high carbohydrate meal like a bagel with cream cheese."

Harry replies. "I've had low blood sugar in the past during dialysis and needed glucose tablets to rescue me."

I respond, "It's possible that dialysis can lead to low blood sugar, but you can use your continuous monitoring device to track the trajectory of your blood sugar during dialysis." We decide to build a bolus insulin scale for Harry at breakfast and see if his blood sugar control improves. We agree to the following scale:

Glucose before breakfast	Insulin dosing
70 – 100	6 units
101-140	10 units

141-180	14 units
181-220	18 units

I tell Harry that we will continue to build foundation stones of care one step at a time and see what his next dataset looks like after implementing this new insulin dosing for breakfast.

Harry asks, "What should I do for lunch and dinner?"

I reply, "Continue with your routine insulin dosing for lunch, dinner, and bedtime until we collect more information." I also interject, "I know you are accustomed to eating a bagel and cream cheese for breakfast before dialysis, but it puts a lot of stress on your body because of the high carbohydrate load. Perhaps you can consider half a bagel and cream cheese plus a protein source for breakfast like a slice of low-sodium turkey or ham." I tell him to consider small tweaks in his diet. I do not push this point, but tell him that changes in diet might also necessitate changing his insulin scale. We move on to Harry's **LIFEMAP** on non-dialysis days. Harry's **LIFEMAP** for non-dialysis days shown below:

LIFEMAP WITH SBGM*
Non-Dialysis Days

Allergies none
Cigarettes none
ETOH none

BB 7:30 a.m.	AB 10:00 a.m.	BL	AL 3:00 p.m.	BD	AD 8:30 p.m.	BT
122	178		262		224	
78	225		148		169	
113	202		193		258	
154	166		251		303	
92	248		182		245	
139	127		236		204	

Wake up ▸ Breakfast ▸ Snack ± ▸ Lunch ▸ Snack ± ▸ Dinner ▸ Snack ± ▸ Bedtime

Time 6:00 7:00 8:00 9:00 10:00 11:00 12:00 noon 1:00 2:00 3:00 4:00 5:00 6:00 7:00 8:00 9:00 10:00 11:00 12:00 mn

Social determinants of disease   Avg. Glucose 188.3 Est. A1C 8.2%

☐ psychosocial ☐ financial ☐ access ☐ substance abuse ☐ compliance
Micro/Macrovascular Complications

[X] Neuropathy ☐ Retinopathy [X] Dialysis ☐ Macroalbumin

☐ CVD ☐ CVA ☐ PVD ☐ Carotid Disease

* **Self blood glucose monitoring**

A similar pattern emerges when his blood sugar is below 100 mg/dl before breakfast. Harry says, "If I don't take insulin before breakfast, I see that the same thing is happening when I don't have dialysis." I say to Harry, "Now you get it. We will use the same scale for your insulin dosing before breakfast on non-dialysis days and see if this works for both." I also tell Harry we cannot use hemoglobin A1C because dialysis causes the number to be falsely low. However, I calculate his average estimated glucose to be 188.3 mg/dl on non-dialysis days and this converts to an estimated A1C of 8.2%. I say, "Your overall glucose is okay at 188, but I think we can improve on this without getting hypoglycemia." I also inform him that if he decided to change his breakfast to a lower carbohydrate, then we would probably need to readjust his insulin dosing scale since he will likely need less insulin before breakfast.

I also ask Harry about his foot and his says it hurts intermittently when he walks. His X-ray results reveal mild osteopenia (decreased bone density) in the feet without evidence of loss of the arch, indicative of possible early Charcot's arthropathy. I send him to the podiatrist for fitting of orthotic inserts into his shoes and a bone scan.

Visit 2 Summary:

1. Interrogate **LIFEMAPs**.
2. Medication reconciliation.
3. Continue SBGM every day with different time; 4 times daily.
4. Referral to podiatry.
5. Return to care in 5-6 weeks with glucose logbook.
6. Visit time is 25 minutes; Level 4 Established Patient visit coding (99214).

Case 8: Visit 3

Harry returns in 6 weeks. He brings in his logbooks for dialysis days and non-dialysis days. In addition, he went to see the podiatrist who performed a bone scan, which revealed a hairline fracture in the arch of the right foot, indicative of early Charcot's arthropathy. He is wearing a boot that takes the weight off his arch and states that he has been in the boot for 2 weeks. He thinks his foot is feeling better already. He states, "My blood sugar readings have improved dramatically since I started take the insulin before breakfast the way you showed me." I look over his logbook and do not need to build a **LIFEMAP** because his post breakfast blood sugar readings are between 108 and 184; a distinct improvement from his previous numbers.

Moreover, he states, "I have had only 1 mild episode of low blood sugar that I corrected with juice." Harry's average estimated glucose is now 155 mg/dl since implementing his new dosing scale before breakfast. This calculates to an estimated A1C of 7.0%, which is perfect. Harry says, "I am still using 10-15 units of Humalog insulin before lunch and dinner and this seems to be working out well. I wish I had known about the **LIFEMAP** twenty years ago. It probably would have prevented me from being here today."

I tell Harry that my goal is to spread the **LIFEMAP** throughout the world so all healthcare providers can learn to properly manage diabetes in a personalized fashion. It will save people from much suffering. I tell Harry he can consider decreasing his **LIFEMAP** reporting to every other day since his blood sugar variance is stable, but that he knows to increase his vigilance if conditions change. Also he has full access to data every day because he is wearing a continuous glucose monitor (CGM). I will see him in follow-up in 3 months.

Visit 3 Summary:

1. Discussion of glucose control.
2. Reinforce continuation of blood glucose monitoring with the **LIFEMAP** approach.
3. Return to care in 3 months with glucose logbook.
4. Visit time 15 minutes; Level 3 Established Patient visit coding (99213).

Case 9
South Asian Diabetes with
Low BMI and Fatty Liver

Case 9: Visit 1

Priya is a 58-year-old Indian female with a history of type 2 diabetes for 10 years. She has a strong family history in both her parents of diabetes and cardiovascular disease. She remembers when she was young in Gujarat there was stress at the dinner table because her mother allotted a portion of rice and her father always ate more, and his blood sugar was uncontrolled after dinner. She is very careful with her diet, but it seems her efforts have not improved her blood sugar control with a recent A1C was 8.7%. Recently she developed right upper quadrant abdominal tenderness and underwent an ultrasound of her abdomen by her primary care physician. This test showed fatty liver disease. She was very upset with this news and asked to see an endocrinologist.

Priya is presently taking liraglutide 1.2 mg daily via subcutaneous injection, metformin 1000 mg twice daily, glimepiride 4 mg, and empagliflozin 10 mg with breakfast. She is also taking lisinopril 5 mg daily, atorvastatin 20 mg daily, and a baby aspirin 81 mg. She states she frequently feels bloated after meals and sometimes needs to take antacids for her stomach. Priya informs me that she works out on a treadmill with a brisk walk 4 times weekly for 30 minutes. She denies chest pain, shortness of breath, numbness in her feet, or frequent low blood sugar. She checks her blood sugar twice daily before breakfast and dinner and self-reports numbers typically between 110-140 mg/dl in the morning and 120-180 before dinner. I tell her that an A1C of 8.7% would translate to an average blood sugar of ~203 mg/dl, so her self-reported blood sugar numbers before meals make sense, but perhaps her blood sugar is higher after food.

I tell Priya, "Diabetes is very prevalent in your family and probably has strong genetics behind the disease. Fatty liver disease is a consequence of insulin resistance in the liver caused by fat molecules that infiltrate the liver tissue and generate inflammation." She says, "But doctor, I am not fat!"

I respond, "Yes, I know that. The term 'fatty liver' is not a good one and it doesn't imply that you are fat. Rather, it is a doctor's way of saying your body accumulates fat stores in the wrong places. It is common in people from Southeast Asia to have this problem with fat storage in the abdominal cavity. This form of fat storage is also associated with cardio-vascular disease. I apologize if you thought I was calling you fat."

On evaluation, blood pressure 130/82, pulse 75, BMI 24.2 (normal).

I tell Priya, "We are going to control your diabetes and parse your cardiovascular risk." I tell her that we are going to start by building a **LIFEMAP**. I ask, "What time do you get up in the morning?"

She tells me, "6:30 a.m."

"And what time is breakfast, 7:00? Tell me the rest of your meal schedule and bedtime." I say.

"Okay," she says. "I eat lunch at 1:00 p.m., dinner at 7:30 p.m., and bedtime is 11:00 p.m."

Based on her **LIFEMAP** we decide on self-blood glucose monitoring at the following times:

SBGM:

Before Breakfast (BB): 6:30 a.m.

After Breakfast (AB): 9:00 a.m.

Before Lunch (BL):

After Lunch (AL): 3:00 p.m.

Before Dinner (BD):

After Dinner (AD): 9:30 p.m.

Bedtime (BT):

Priya says, "That's a lot of blood sugar testing." I respond, "You only need to test three days per week, four times daily in the beginning. We build the **LIFEMAP** with data obtained at home and then use trend analysis to build a treatment plan. The trend analysis assumes we are viewing blood sugar results on an average day in your life and the ups and downs even out over time." Priya understands and says she will start checking her blood sugars as discussed.

She asks, "What do I do on the days where I don't check my blood sugar?" I respond, "If you feel bad or think your blood sugar is low or out of control, then certainly check you sugar. However, for the purpose of the **LIFEMAP** it isn't necessary to check your blood sugar at other times. You will see some really bad numbers, especially when you check your sugar 2 hours after meals. In my world, all numbers are good numbers. The only numbers that are bad numbers are no numbers!"

I say, "When you return for your next visit we will de-convolute your **LIFEMAP** and you will see how easy and logical it is to get your blood sugar controlled." Finally, I state, "Your family risk of heart disease is very high and you are taking medicine for both diabetes and cholesterol. I like to use a simple test called a coronary calcium scan that gives good information about the general health of the arteries in your heart. If the calcium score is high, then it is worth doing more tests to prevent a heart attack. I will give you a prescription for this test."

Medication List:
- ✓ liraglutide 1.2 mg subcutaneous daily
- ✓ metformin 1000 mg twice daily before meals
- ✓ empagliflozin 10 mg daily
- ✓ glimepiride 4 mg daily
- ✓ aspirin 81 mg daily
- ✓ atorvastatin 20 mg daily at bedtime
- ✓ lisinopril 5 mg daily

Visit 1 Summary:
1. Build **LIFEMAP**.
2. Medication reconciliation.
3. Obtain SBGM 3 days per week, 4 times daily.
4. Obtain coronary calcium scan.

5. Return to care in 4-5 weeks with glucose logbook.

6. Visit time is 45 minutes; Level 4 New Patient visit coding (99204).

Case 9: Visit 2

Priya returns in 4 weeks with a sour look on her face and a logbook with neatly written glucose values as previously discussed. She says, "My blood sugar is really bad and my heart is bad." I respond, "Let's see what your numbers look like," and she shows me her LIFEMAP below:

LIFEMAP WITH SBGM*

Allergies none
Cigarettes none
ETOH none

BB 6:30 a.m.	AB 9:30 a.m.	BL	AL 3:00 p.m.	BD	AD 9:30 p.m.	BT
177	234		205		158	
163	194		188		221	
188	246		202		195	
142	179		193		180	
166	205		241		218	
172	213		199		215	

Wake up ▸ Breakfast ▸ Snack ± ▸ Lunch ▸ Snack ± ▸ Dinner ▸ Snack ± ▸ Bedtime ▸

Time 6:00 7:00 8:00 9:00 10:00 11:00 12:00 noon 1:00 2:00 3:00 4:00 5:00 6:00 7:00 8:00 9:00 10:00 11:00 12:00 mn

Social determinants of disease   Avg. Glucose 195.6 Est. A1C 8.4%

☐ psychosocial ☐ financial ☐ access ☐ substance abuse ☐ compliance
Micro/Macrovascular Complications

☐ Neuropathy ☐ Retinopathy ☐ Microalbumin ☐ Macroalbumin

☐ CVD   ☐ CVA   ☐ PVD   ☐ Carotid Disease

* Self blood glucose monitoring

She also has a report for her coronary calcium score that shows the following:

Extensive coronary artery calcifications throughout the coronary arteries. Small calcium plaque in left main coronary artery (score 2.5). A large calcific plaque is present in the proximal, mid, and distal LAD (score 879.3). Multiple calcific plaques throughout the right coronary artery (score 199.5). Total coronary calcium score is 1081.3. This score is in the 90th percentile and comes with substantial risk of significant coronary artery disease in at least one vessel.

I explain to Priya that our highest priority is to get her heart evaluated and then we can work on her blood sugar management after she is safe. I immediately refer her to a cardiologist for angiography and possible stent placement if needed.

I also explain to Priya that her LIFEMAP shows liver insulin resistance with high-fasting blood sugar in the morning. I tell her we can get her blood sugar under better control in the morning, which will lead to better sugar control throughout the day. I also let her know that sometimes the blood sugar improves after heart treatment because of improved blood flow throughout the body. I tell her I am going to change her liraglutide shot once daily to a different shot at bedtime called Soliqua, which combines a drug similar to liraglutide with long-acting insulin. This combination will lead to better fasting blood sugar in the morning and additionally give her better control throughout the day. I ask that she continue with glucose monitoring 3 days per week as before.

Visit 2 Summary:
1. Referral to cardiology for coronary artery evaluation.
2. Discontinue liraglutide.
3. Begin Soliqua 25 units at bedtime.
4. Continue with SBGM 3 days per week, 4 times daily.
5. RTC in 6-8 weeks after heart evaluation.
6. Bring in logbook on next visit.
7. Visit time is 25 minutes; Level 4 follow-up visit coding (99214).

Case 9: Visit 3

Priya returns in eight weeks. She says, "Doctor Bleich, you saved my life. Thank you." She continues, "The cardiologist said your endocrinologist is very smart and he saved me from a major heart attack!" Priya states, "I now have two stents keeping my arteries open and I actually feel better." Priya is now taking a blood thinner called clopidogrel along with her baby aspirin to keep her stents in good shape. She also says, "My blood sugar is good now," and shows me her most recent glucose log from the past two weeks.

LIFEMAP WITH SBGM*

Allergies none
Cigarettes none
ETOH none

BB 6:30 a.m.	AB 9:30 a.m.	BL	AL 3:00 p.m.	BD	AD 9:30 p.m.	BT
123	148		155		137	
114	128		138		129	
108	135		147		172	
127	144		138		146	
103	115		129		126	
98	128		154		158	

Wake up ➤ Breakfast ➤ Snack ± ➤ Lunch ➤ Snack ± ➤ Dinner ➤ Snack ± ➤ Bedtime ➤

Time 6:00 7:00 8:00 9:00 10:00 11:00 12:00 noon 1:00 2:00 3:00 4:00 5:00 6:00 7:00 8:00 9:00 10:00 11:00 12:00 mn

Social determinants of disease   Avg. Glucose 129.2 Est. A1C 6.1%

☐ psychosocial ☐ financial ☐ access ☐ substance abuse ☐ compliance
Micro/Macrovascular Complications
☐ Neuropathy ☐ Retinopathy ☐ Microalbumin ☐ Macroalbumin
☒ CVD ☐ CVA ☐ PVD ☐ Carotid Disease
* Self blood glucose monitoring

Her average glucose is now 129.2 mg/dl, which converts to an estimated A1C of 6.1%. Priya states, "I increased Soliqua at bedtime to 34 units and my fasting blood sugar has been perfect." I tell Priya that she is on an excellent diabetes drug regimen and now that her heart is stable she should be good to go for many years. I reinforce that the **LIFEMAP** is a critical tool to use going forward because it will tell her when her blood sugar is getting out of control. I add coronary vascular disease (CVD) to her **LIFEMAP**. I say, "It would be good to continue checking your blood sugar 1-2 days per week, 4 times daily, forever. I will see you back in 3-4 months with new lab tests so we can make certain everything is in place.

Visit 3 Summary:

1. Interrogate **LIFEMAP** and add CVD.
2. Medication reconciliation.
3. Reinforce need for continued blood sugar monitoring 1-2 times weekly, 4 times daily.
4. Obtain labs prior to next visit: A1C, lipid panel, CMP, CBC, urine microalbumin to creatinine ratio.
5. Return to care in 3-4 months.
6. Bring in glucose log.
7. Visit time is 15 minutes; Level 3 follow-up visit coding (99213).

Chapter 13

LIFEMAP in Motion

The Diabetes LIFEMAP is a disruption to the day-to-day practice of chronic diabetes care. It requires both healthcare provider and patient to buy into 50:50 relationships for diabetes management: the provider must develop a tailored roadmap for success that fits into the lifestyle of the patient (the LIFEMAP); and the patient must agree to provide information to the healthcare provider or system and try to improve lifestyle. It also requires flexibility in the healthcare system and electronic health record (EHR) to accommodate the individual quirks and differences in how individuals get through their day rather than using standardized EHR templates that are not user friendly and eat up valuable visit time.

In our time-constrained world of medicine, it is critical to prioritize those interventions that impact the well-being of the patient and those that don't. The LIFEMAP puts focus on controlling blood sugar since this is why the patient came to your office. The LIFEMAP also recognizes that diabetes care is happening 24/7 and that most of this time is not spent in the doctor's office.

Therefore, we have adapted the LIFEMAP to use smartphone technology to give patients access to the healthcare system and vice versa. Our cloud-based LIFEMAP will not only give patient and provider portals but is also designed to proactively message and collect data from the patient at their preferred times of day based on their lifestyle, thereby increasing the precision, accuracy, and usefulness of the data to make informed care decisions. In the near future, I envision a chronic care delivery system that includes office visits as well as telemedicine interactions to maintain well-controlled blood sugar. The LIFEMAP will provide an integral service for patients to facilitate efficient, high-quality diabetes care using telemedicine modalities. Our future for chronic care delivery will embrace cost-saving technologies that improve outcomes. The LIFEMAP is positioned to play a lead role in this transformation of diabetes care.

Social determinants of disease matter a lot because it is often these "life gets in the way" events that upend the best plans. Simple thing like getting to the doctor's office or healthcare facility can be barriers to care for many individuals. Moreover, many of the activities that take place in the chronic care setting are goal setting, care maintenance, medication refills, and disease screening. Most of these activities do not require an office visit, yet we are stuck with an old model of care delivery that does not work for the 21st century.

We have previously discussed I organically derived the LIFEMAP; I never had intention to reinvent the chronic healthcare delivery system for diabetes care. I am humbly appreciative to all my patients with diabetes who taught me the principles elucidated in this book. My patients helped me to realize that diabetes care needs a carefully constructed plan that is integrated with lifestyle and day-to-day activities. My patients got me to realize that a personalized plan for care is practical and goes a long way to empowering individuals in their care management.

When I first confronted my wonderful patients in Newark, New Jersey, I felt overwhelmed by their burden of disease, social injustice, poverty, language barriers and lack of understanding about diabetes. Over time, it became clear to me that quality diabetes care is in part about teaching patients why certain "doctor things" are important. I also learned that respecting limitations and encouraging patients not to give up is essential to their well-being. Since I run the diabetes program at University Hospital in Newark, I see more complex patients who have already encountered numerous healthcare providers who have failed to improve their blood sugar control. These individuals challenged me to find a road forward for them. In fairness, I could not ask my patients to keep trying if I was willing to give up. This attitude plus a lot of careful listening led me to develop the LIFEMAP. Now, I receive routine comments from my diabetes patients like, "Why didn't someone tell me this before?" or "I wish I had known this years ago because now I understand what I am doing." The LIFEMAP simplifies and clarifies a complex disease that is overwhelming to patients. I have both positive and negative feeling towards these comments. I am pleased that I bring new understand and hope to my patient with diabetes, yet I am sad it took me this long to understand and develop the LIFEMAP.

We think of diabetes as a binary disease; you have it or you don't. This is an inaccurate view because diabetes follows a trajectory for most individuals. As blood sugar rises, glucose toxicity causes injury to vital tissues that are important for maintaining glucose control. More so, as we age our metabolic rate decreases and our ability to utilize fuel decreases. This becomes a setup for worsening blood sugar control. As insulin production wanes over time, individuals with diabetes often transition from a disease controlled with pills to one that requires insulin shots. This transition is often unrecognized early in the disease course and catches many patients by surprise. The dreaded insulin shot, while life changing in many ways, is life saving for those who would otherwise end up in the hospital with uncontrolled hyperglycemia and dire consequences.

The case examples presented in chapter 12 are amalgamations of scenarios I deal with routinely using the LIFEMAP. These scenarios appear easily resolved because of the LIFEMAP, but I was lost without it.

In Case 1, Juan suffers from uncontrolled type 2 diabetes and diabetic neuropathy. He drinks a six-pack of beer on the weekends, which might add to his diabetic neurotoxicity and increases his carbohydrate consumption. The LIFEMAP identifies a pattern of uncontrolled blood sugar that is treatable with an SGLT-2 inhibitor and dietary modification because his post meal blood sugars are all consistently high, but his fasting blood sugars are lower. He is successful because he cuts back on his beer consumption, decreases his carbohydrates at mealtimes, and takes a new medication that causes glucose loss in the urine. His neuropathy improves because he lowers his blood sugar and decreases alcohol-induced stress on his nerves. The LIFEMAP gives Juan control over his blood sugar and weekly feedback on his progress.

In Case 2, Shauna's life is complicated because she is a single parent with two young children and uncontrolled blood sugar. She wants to maintain her health and is receptive to my recommendations. We work together to determine an optimal treatment strategy with help from the LIFEMAP. She is ultimately able to achieve success with a combination drug at bedtime that gets her fasting blood sugar controlled with blood sugar monitoring that is tailored to her lifestyle and work schedule.

In Case 3, Jeff has had poorly controlled long-term type 2 diabetes that caused microvascular complications of diabetic retinopathy (eye disease) and neuropathy (nerve disease). He is on two shots of insulin daily plus metformin and does not know what the road forward looks like for better blood sugar control. We identify that he also has kidney disease that could ultimately lead to end-stage kidney disease and dialysis. Jeff is eager to improve his blood sugar control but has no compass for how to achieve this goal since his blood sugar management has been the same for many years. As a first step towards better blood sugar control, Jeff's LIFEMAP teaches us that we can use an unconventional approach by adding a shot of long-acting insulin at bedtime to his two shots of mixed insulin before breakfast and dinner. This leads to improvement in fasting blood sugar but he continues to have elevated blood sugar after meals.

We decide to add a sugar-blocking drug called acarbose and this smooths out his LIFEMAP on the following visit. However, Jeff still needs additional glucose control. We decide to cautiously try an SGLT-2 inhibitor to see if his kidneys tolerate the medication and whether this does the trick to get his blood sugar nicely controlled. When Jeff returns his LIFEMAP shows excellent glucose control without kidney side effects. Jeff ends up on an unconventional treatment plan with one shot of mixed insulin before breakfast and one shot before dinner, a shot of long-acting insulin at bedtime, acarbose before meals to decrease carbohydrate absorption, and SGLT-2 inhibitor to offload glucose into the urine. The LIFEMAP led us to this treatment strategy without bias.

In Case 4, Jenna teaches us about the problem with fixed-dose long-acting insulin at bedtime. Pharmaceutical companies have marketed their long-acting basal insulins as "fixed dose" solutions to uncontrolled hyperglycemia because this is an easy approach to dosing insulin. When used at bedtime, long-acting insulin analogs control hepatic glucose production and fasting blood sugar overnight. However, a problem inherently arises in some individuals who have variable glucose levels at bedtime, good tissue sensitivity to insulin, and don't check their blood sugar routinely before bedtime. For such individuals, low blood sugar overnight or in the early morning becomes a noticeable risk.

The LIFEMAP easily resolves this problem and provides a real-world solution. For Jenna, building a basal bedtime insulin dosing scale becomes a life saver because it fixes her problem of overnight hypoglycemia. Jenna is happy to test her blood sugar every night before bedtime if she can go to sleep without worrying that her insulin dose will cause low blood sugar when she is sleeping. Please note that in a few select patients with very labile blood sugar (typically with type 1 diabetes), even the LIFEMAP cannot resolve all hypoglycemia.

Case 5 is unfortunately one of my more common scenarios. Grace is tough to manage because she has obesity, type 2 diabetes, and severe insulin resistance. The dilemma of treating Grace with insulin, especially at high doses, is that it causes weight gain. She is locked in because we need to get her blood sugar controlled at the expense of further weight gain. I switch Grace to a highly concentrated form of insulin called U-500 because it delivers high-dose insulin in a small volume of liquid and improves glucose control in many patients. Grace has already enrolled herself to have weight-loss (bariatric) surgery so our goal becomes somewhat easier since we do not have to worry as much about pre-surgical weight gain. Most important, hemoglobin A1C >8.0% significantly increases surgical risk, including infection, problems with wound healing, and overall mortality. Therefore, we work together to get her A1C < 8.0% prior to gastric bypass surgery. After her surgical procedure Grace will have ~70% chance of no diabetes 2 years after surgery and ~30% chance of no diabetes 10 years after surgery.[50] A surgical option for Grace is the right choice and will give her better quality of life and blood sugar control.

In Case 6, we get to use the LIFEMAP twice because Martha is a shift worker. Her schedules differ significantly on those nights when she works and those days when she is off, thereby necessitating two separate LIFEMAPs to fully capture her blood sugar variance. My experience with shift workers is that, as a group, their diabetes is difficult to manage

50   L. Sjostrom, "Review of the key results from the Swedish Obese Subjects (SOS) trial – a prospective controlled intervention study of bariatric surgery," *Journal of Internal Medicine*, 2013, 273(3):219-234.

because they develop severe insulin resistance from sleep deprivation and changing meal patterns. For Martha, we find a compromise that keeps her blood sugar better controlled than without the LIFEMAP, but not quite optimal. In my view, shift work, with rare exception, falls under a category of social determinants of disease because most individuals would prefer not to do it if they had a choice. Usually, the night shift is less desirable and therefore pays more money and is harder to fill than the day shift. Over many years shift workers adjust to the flipped day-night schedule and find it difficult to revert to a "normal" schedule when the opportunity arises. For Martha, we create two different treatment plans to accommodate her difficult lifestyle.

Case 7 deals with Patricia, who might suffer from alcoholism, and is leading a very stressful life. She understands that diabetes is serious and can cause her harm but cannot find a way to fit it into her life. We decide that a continuous glucose monitor will give us a snapshot of her blood sugar control for several weeks. She is willing to follow my instruction but is not forthcoming with details about her psychosocial stresses. I work around the edges with Patricia to establish my role as helper, not judge. She understands that I expect her to meet me halfway for her care plan to work and we find a way forward with the LIFEMAP to get her blood sugar under better control. Her life stresses are still there, but here it was beneficial to take what the patient can give rather than to impose my will on a patient. The latter approach inevitably leads to failure. Patricia recognizes that I am trying to build a trust relationship that focuses on her diabetes only. We accomplish the goal of better blood sugar control and relieve one of Patricia's problems.

Case 8 is complex because Harry is on dialysis and therefore needs two LIFEMAPs. As it turns out, his blood sugar patterns are similar on the two LIFEMAPS, so we consolidate the glucose analysis but maintain a day-specific timing for insulin dosing. Also, Harry has the common problem of not taking insulin before meals when his blood sugar is low.

The situation arises because fixed-dose insulin does not work before breakfast when your blood sugar is 92 mg/dl on Monday and 158 mg/dl on Tuesday. Harry recognized that his usual fixed-dose insulin before breakfast when he is at 92 will cause hypoglycemia, so instead of using a lower dose, he opts out and does not use insulin at all. This results in post breakfast hyperglycemia that is easily viewed on his LIFEMAP. A simple bolus dosing scale before breakfast corrects this problem and puts Harry on track for excellent blood sugar control.

In *Case 9*, we learn about Priya. She has a strong family history of diabetes and heart disease and I pivot on her LIFEMAP because identifying coronary artery disease is a high-impact intervention that can be fatal if overlooked. Rather than focusing my attention primarily on her glucose control I recognize that she is a high risk for unsuspected coronary artery disease. Individuals with diabetes often do not display classic symptoms of heart disease like shortness of breath on exertion or chest pressure on activity. Often, symptoms are subtle, like frequent indigestion, fatigue, or nothing. A coronary calcium scan is an excellent screening tool, which is underutilized for individuals with diabetes.[51] Priya has a very high calcium score and subsequent coronary angiography reveals significant blockage in two vessels. She has stents placed in two blood vessels to prevent a heart attack. Moreover, we used the LIFEMAP to improve her blood sugar control with a combination of two drugs in one bedtime injection. The LIFEMAP helps to focus the visit on high-impact interventions that sometimes can save a life.

The nine cases presented here describe common scenarios seen by many healthcare providers and experienced by many patients. The LIFEMAP is also a useful tool for the ~85 million Americans and ~350 million individuals worldwide who suffer from prediabetes. Since a significant

---

51 S. Agarwal, A. J. Cox, D. M. Herrington, et al., "Coronary Calcium Score Predicts Cardiovascular Mortality in Diabetes: Diabetes Heart Study," *Diabetes Care*, 2013, 36(4):972-979.

percentage of these individuals will transition to overt diabetes at some time point, the LIFEMAP gives them a way to monitor and hopefully preempt this transition. The LIFEMAP provides a standardized approach to diabetes that integrates lifestyle factors, blood sugar monitoring, and treatment options. In the near future, our cloud-based LIFEMAP smartphone platform will transform diabetes care worldwide. It is my hope that among these nine cases you will find yourself if you are the patient. For the healthcare provider, we often do not know what the road forward looks like for a patient, so we fall back on old habits. I know that before the LIFEMAP, I too fell back on old habits to treat my patients with diabetes. I only knew what I had been taught. Now the LIFEMAP carries me forward with my patient care by clarifying optimal treatments, understanding limitations as human beings, and using data in a thoughtful way to help make decisions.

Conclusion

The LIFEMAP is a structured approach to diabetes care that relies on numbers from self-blood glucose monitoring at home. In reality, this is where diabetes care is most important. The LIFEMAP transforms diabetes care from an older model that is office based and physician centric to a 21st century model that is home based, patient centered, and data driven. New technology plays an important role in this transformation. The reason why the LIFEMAP is so powerful is simply because it is a streamlined attempt to recapitulate the blood sugar dynamics of healthy physiology in individuals who do not have diabetes. It is rigorous because it transforms an antiquated model of generalized top-down care and treatment to a personalized bottom-up approach that embraces out-of-office care and telemedicine.

Modern life is complex. As a patient once said to me, "Diabetes is a full-time job that you never applied for and don't get paid." Unfortunately, this statement has stayed with me for more than 20 years because it cuts to the truth. With this understanding, the LIFEMAP is my best attempt to make diabetes care a little less onerous for those who struggle with controlling their blood sugar every day.

We are actively working on the mobile version of the LIFEMAP, through our cloud-based diabetes management platform. Once fully implemented, the LIFEMAP will enable personalized diabetes care using digital patient mobile technology (smartphones) that will inform a cloud-based data warehouse and prescribing algorithms. Both patient and provider will benefit from a bidirectional telemedicine care management platform that will enable better quality and improved access to personalized, efficient diabetes management. Artificial intelligence (machine learning) will transform the data we acquire into treatment options that are unbiased and data-driven. This will allow primary care providers and other healthcare experts to deliver optimal, personalized diabetes care to their patients.

Technology alone cannot overcome the many hurdles we face in life. Social determinants of disease (SDD) are an overarching phrase that encompasses many different challenges people face every day. Cigarette smoking, alcoholism, food addiction, drug addiction, mental illness, poverty, housing problems, emotional stress, health insurance problems, and access to healthcare are just a few of these hurdles we overcome to lead healthy lives. Our healthcare system fails individuals in many of these aspects. It does not provide holistic answers to the difficult problems that plague humanity.

The LIFEMAP is for diabetes care, but it does not solve these bigger problems. The LIFEMAP has a special focus on capturing information about social determinants of disease because it is well known that quality diabetes care cannot be achieved when bigger social hurdles are in play. One might say, "There is no room for a second unpaid job like diabetes management when you already work two jobs to keep afloat."

However, our aspiration in cataloging SDD with the LIFEMAP will allow healthcare providers, researchers, and government agencies to reallocate resources to problems that interfere with diabetes care. In turn, this might produce better care at the root cause of disease and better resource allocation. Since dollars spent on depression (for example) might yield better quality of life for those who so suffer, it might also improve diabetes care and lower cost overall if the outcome is improved mental health and better glucose control. While this hypothesis is only speculation at present, the LIFEMAP will allow healthcare providers and healthcare systems to take a deep dive into the morass of quality, efficiency, and cost of healthcare. We anticipate using our big data source achieved through the LIFEMAP to answer questions that are still knowledge gaps today.

The LIFEMAP is intended for diabetes care in its present formulation that can be scripted into medical charts by healthcare providers, but we anticipate transforming chronic care management broadly for both healthcare provider and patients. As we previously discussed, the rubric for acute disease management cannot be sustained in the time-constrained chronic care environment because it mixes what I call "high-impact interventions" with "moderate- and low-impact interventions." This quirk evolved over decades in the practice of medicine as an unintended consequence of translating the acute healthcare model into the chronic care setting without really understanding that these two models are entirely different. This "goof" has had significant consequences for both parties; the healthcare provider spends time on low-impact items because there are regulations in place that require accountability for what is done at the office visit, and the patient gets suboptimal care.

Such items as routine heart, lung, and abdominal examinations are very low-yield activities for individuals without specific complaints. Yet they are still beans that get counted when the provider submits a billing claim to the health insurance provider. Low-impact interventions steal time away from high-impact interventions such as establishing a doctor-patient relationship by listening to a heartbreaking story of a loved one who passed away, hearing about a homeless individual who is struggling to find housing in a shelter, or empathizing with a well-to-do family whose daughter was recently admitted to a rehab center for opioid addiction. Getting to diabetes and glucose management needs to be placed in the context of a suffering human being. It is my hope that the LIFEMAP will make things easier for individuals with diabetes and provide a way forward in a complex, time-constrained world, because nobody wants a full-time second job that you don't get paid for.

As our technology continues to evolve, many aspects of diabetes care traditionally performed in a doctor's office can now be delivered through telemedicine. Even medication refill requests can be done electronically through modern healthcare records. The government and the Center for Medicare and Medicaid Services (CMS) has recognized that telemedicine will contribute increasingly to healthcare delivery and therefore it has set up "billing" codes for healthcare providers so they can get paid for their time. At present, rules and regulations for telemedicine interactions require real-time audio and visual communication between patient and healthcare provider, but this is changing. Specific diseases are designated for eligible services and we anticipate transforming routine diabetes care into a telemedicine intensive process using the LIFEMAP. This will allow more frequent patient access to the healthcare system, improved efficiency for the care provider, and better outcomes.

For the healthcare provider, the ability to obtain a dataset and have analytic tools to provide best practice recommendations that are personalized is easy with our technology. This approach to diabetes care makes all the sense in the world because diabetes is a data-driven disease. If we remove our medication bias as healthcare providers and ask the question, "What is most important for an individual with diabetes?", then the answer is always, "Control the blood sugar." We can achieve this goal in many ways with many different medications and lifestyle interventions. But it is fundamentally different for each person. The LIFEMAP transforms diabetes care into a personalized, unbiased approach to blood sugar control and provides a new understanding of diabetes management for the future.

We hope the LIFEMAP will solve some of the challenges for patients with diabetes and make life better. Access to care is one of the social determinants of disease that can be overcome with the LIFEMAP. Other barriers to care will also be overcome once the full force of the LIFEMAP is in place. This will be a good day for medicine.

About the Author

Dr. Bleich is Chief, Division of Endocrinology, Diabetes, & Metabolism and Professor of Medicine at Rutgers New Jersey Medical School in Newark, N.J. He is a nationally recognized expert in both type 1 and type 2 diabetes and has played leadership roles in diabetes organizations including the American Diabetes Association, the Juvenile Diabetes Research Foundation International, and the Department of Defense Diabetes Grant Review. Dr. Bleich has conducted grant-funded bench research on pancreatic beta-cell function and has spent >10,000 hours caring for individuals with diabetes, as well as mentoring and teaching medical students, residents and endocrine fellows. He is presently focused on improving the quality of diabetes care and developing a data-driven holistic approach to patient care. Dr. Bleich received his MD degree from New York Medical College, internal medicine training from Maimonides Medical Center in Brooklyn, NY, and fellowship training from Harvard Medical School. He lives in Westfield, N.J.